SpringerBriefs in Molecular Science

Green Chemistry for Sustainability

W0080293

Series editor

Sanjay K. Sharma, Jaipur, India

More information about this series at http://www.springer.com/series/10045

Ying Li · Anne-Sylvie Fabiano-Tixier
Farid Chemat

Essential Oils as Reagents in Green Chemistry

 Springer

Ying Li
Anne-Sylvie Fabiano-Tixier
Farid Chemat
Université d'Avignon et des Pays de
  Vaucluse
INRA-SQPOV, UMR408
GREEN Extraction Team 84000
Avignon
France

ISSN 2212-9898
ISBN 978-3-319-08448-0   ISBN 978-3-319-08449-7   (eBook)
DOI 10.1007/978-3-319-08449-7

Library of Congress Control Number: 2014944544

Springer Cham Heidelberg New York Dordrecht London

Printed on acid-free paper

Springer is part of Springer Science+Business Media (www.springer.com)

Preface

The essential oil is one of the most promising themes that can strongly contribute to the Green Chemistry, not only in research laboratories, but also in various industries and at the teaching level from primary schools to universities. This conclusion was based on two observations. Essential oils are widely used in perfume, cosmetic, pharmaceutical, agricultural, and food industries. It has long been recognized since antiquity to possess biological activities, including antibacterial, antifungal, antiviral, antimycotic, antitoxigenic, antiparasitic and insecticidal properties. A large number of essential oils and their constituents have been investigated for their antioxidant properties in cooked and fresh food products. In recent years, researchers and industries are more focused on the major compounds of essential oils in order to use them as bio-based solvents for extracting valuable metabolites (e.g., fat and lipid, carotenoids, polyphenols) or as reagents (synthons) for newly bio-based chemicals for pharmaceutical, food or cosmetic purposes.

As a main difference from previously published books in this area, readers like chemists in synthesis or analysis, biochemists, chemical engineers, physicians, food and agro- technologists will find a deep and complete perspective regarding essential oils. Following an introduction to the history of essential oils (Chap. 1), Chap. 2 details conventional and innovative extraction techniques. Biological applications in which essential oils have afforded spectacular results are discussed extensively in terms of fundamentals, tests, and applications: antioxidants (Chap. 3), antimicrobials (Chap. 4), and insecticides (Chap. 5). The last two chapters give new directions for research and industry by using major or single components in essential oils as bio-based solvents (Chap. 6) or as green reagents for syntheses (Chap. 7).

We wish to thank sincerely all our colleagues from "GREEN Extraction Team" in Avignon University who have collaborated in essential oil's applications. We express our thanks to the personnel from Springer who have offered their time and support, especially Dr. Sonia Ojo for her help to make this SpringerBrief possible. On the other hand, we are totally convinced that this book is the starting point for future collaborations in new "green chemistry of essential oils" between research, industry and education.

Contents

About the Authors

Ying Li is a Doctoral Researcher of Natural Product Chemistry from the GREEN Extraction Team, Avignon University in France. He received his Master of Green Chemistry as specialization from University of Toulouse. His current research interest is the integration of innovative techniques (e.g., ultrasound and microwave) and agro-based solvents (e.g., vegetable oils, terpenes, etc.) for green extraction of bioactive compounds from natural bioresources, which leads to greener processing procedures and novel value-added end products with great potential in food, cosmetics, nutraceutical, and pharmaceutical industries.

Anne-Sylvie Fabiano-Tixier is Associate Professor of Chemistry in GREEN Extraction Team at Avignon University (France) and ORTESA LabCom Naturex-UAPV. She holds a Ph.D. degree in Chemistry of Biomolecules from the University of Toulouse. Her expertise is extraction techniques (especially microwave, ultrasound, and green solvents) and antioxidant activity applied to food, pharmaceutical, and cosmetic domains. More than 25 scientific peer-reviewed papers, three patents, and 50 communications in events document her research activity.

 **Farid Chemat** is a full Professor of Chemistry at Avignon University, Director of GREEN Extraction Team (alternative techniques and solvents), Codirector of ORTESA LabCom research unit Naturex-UAPV, and scientific coordinator of "France Eco-Extraction" dealing with dissemination of research and education on green extraction technologies. His main research interests are focused on innovative and sustainable extraction techniques, protocols, and solvents (especially microwave, ultrasound, and bio-based solvents) for food, pharmaceutical, fine chemistry, biofuel, and cosmetic applications. His research activity is documented by more than 140 scientific peer-reviewed papers, nine books, and seven patents.

Corresponding address

Université d'Avignon et des Pays de Vaucluse, INRA, UMR408, GREEN Team Extraction, 84000 Avignon, France

farid.chemat@univ-avignon.fr

http://green.univ-avignon.fr/

Chapter 1
History, Localization and Chemical Compositions

Abstract A brief introduction of essential oils (EOs) is given, including their historical uses, definition and localization in various plant organs. The EOs are then described in terms of their organoleptic characteristics, physical properties, chemical compositions and relevant chromatographic profiles. Besides, intrinsic and extrinsic parameters influencing the quality and the composition of EOs are discussed.

Keywords Essential oils · History · Definition · Composition · Localization · Variability

1.1 History

The use of fragrances and aromas dates back to ancient civilizations when fragrances have firstly played an important role in nutritional, aesthetic and spiritual applications during pharaonic era (Fig. 1.1). Specific aroma compounds such as sesquiterpenes in the frankincense extracts, which have been extensively used for religious rituals, were discovered in bandages of mummies.

Natural aromatic plants and gums (e.g. olibanum, myrrh, galbanum) which are macerated in vegetable oils or for medical purposes in water, were widely traded in India, Persia and Egypt. Greeks and Romans have made their contribution on the establishment of distillation fundamentals such as "ambix" invention which can be applied for both extraction and distillation. They also found a simple fumigation way in sickrooms through burning lavender twigs, which was used in the northern Europe as well for plague prevention. The Muslim civilisation strongly promoted the development of the spice trade, and the distillation and extraction techniques afterwards. Geber (712–815) and Avicenne (930–1037) invented the alembic, distillation and refrigerant, respectively. Since the initial preparation of distilled water in medieval times, the use of essential oils (EOs) has experienced the empirical application of specific aromatic plants (e.g. clove and nutmeg) in Renaissance. The Florentine flask invented by Giovanni Battista Della Porta (1533–1615) has

Y. Li et al., *Essential Oils as Reagents in Green Chemistry*, SpringerBriefs in Green Chemistry for Sustainability, DOI 10.1007/978-3-319-08449-7_1

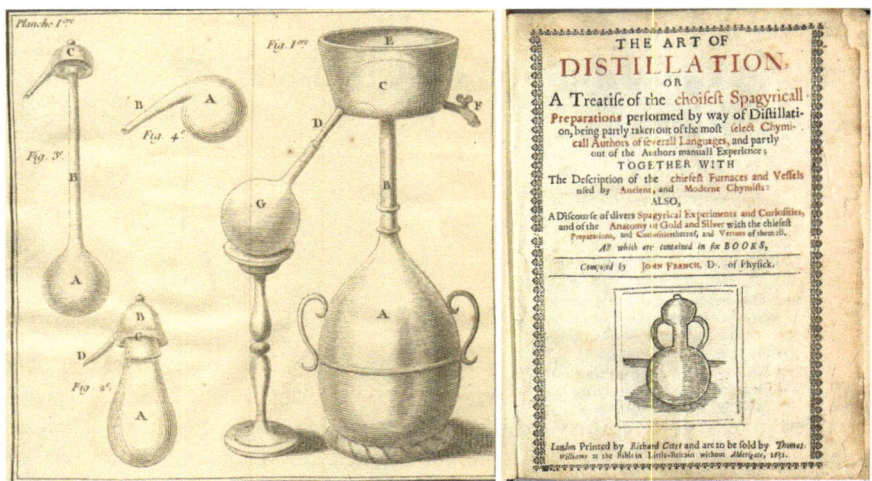

Fig. 1.1   Alembic for distillation of rose water (Macquer 1756; French 1651)

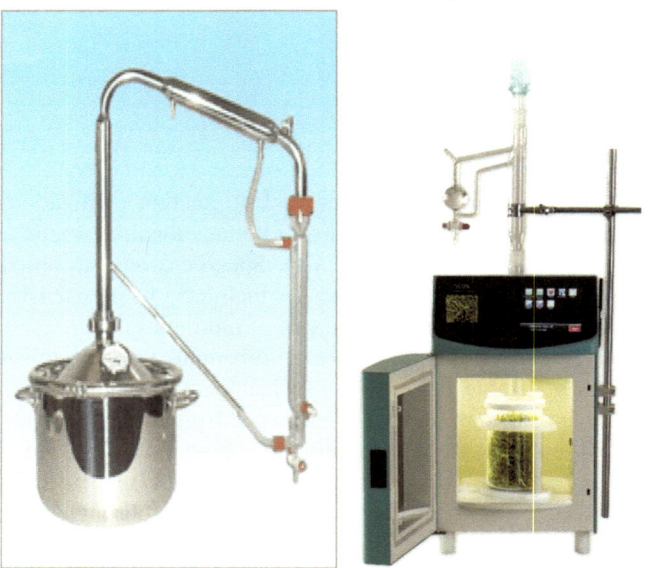

Fig. 1.2   Conventional and innovative (microwave) Clevenger for extraction of essential oils

facilitated the separation of EOs and water after the extraction or distillation of EOs (Fig. 1.2). Since then, the aroma industry has experienced the popularity of leather fragrances in 16th century, the creation of bergamot and neroli oil in 17th century, the spread and development of the first synthetic aromatics and fern fragrances in 19th century, the acclaimed perfumers and famous fragrances in

20th century and the boom of their use in the field of perfumes, cosmetics, food pharmaceutical and household nowadays, as well as their increasingly important impact on aroma-, psycho- and physio-therapy (Smadja 2009).

EOs are odoriferous oily liquids obtained from all parts of plants through volatilization without decomposition by physical methods only like water or hydrodistillation of spices and aromatic herbs or squeezing citrus peels (Fig. 1.3). As the result of their exclusive odour, they are universally and empirically used in a direct or indirect way to help attract or repel insects for pollination or to prevent parasites or herbivores in most cases. The aroma compounds resulting from plant metabolism are formed and stored in one or several plant organs. Are EOs beneficial or harmful to human and environment? Can they be applied in green chemistry? More studies have been urged to explore their functionality so as to get a complete knowledge for using EOs as green reagents in future sustainable strategies.

Fig. 1.3 Industrial extraction of essential oils in north and south countries

1.2 Definition and Localization

The definition of EOs has been attempted by numerous studies or organizations, from which the French Association for Standardization (AFNOR) proposed a definition in 1998 corresponding to the following revised ISO 9235 in 2013. The essential oil is the product obtained from a raw material of plant origin by means of either distillation or mechanical processes for *Citrus* only, or by "dry" distillation for woods. The essential oil is then separated from the aqueous phase by physical methods (ISO 9235).

The term "essence" defines the exhaled fragrances that are natural secretions produced by different plant organs. These fragrances are due to the presence of volatile aroma compounds in plant cells. The term "oil" denotes the lipophilic (i.e., hydrophobic) and viscous nature of these substances while the term "essential" signifies their preciousness and typical fragrance of plants. It is also important to notice the difference between aromatic extracts and EOs. An aromatic extract is usually composed by at least one aromatic substance extracted from a plant or parts of animals using various methods with organic solvents such as hexane or ethanol. However, only exogenous or endogenous water is used as solvent to obtain an essential oil by means of various separation techniques such as hydro-distillation and diffusion, etc.

EOs can be found in various organs of plants from different families, of which Lamiaceae, Myrtaceae (*Eucalytus*) and Rutaceae (*Citrus*) families are found to have a high level of volatile aroma compounds. In fact, only 10 % of plant species can synthesize and secrete small amounts of essence and they are thus called aromatic plants. Moreover, the EOs extracted from different organs in the same plants are variable for their names and uses because of their odours (Chemat et al. 2013). EOs are stored in the cytoplasm of certain secretory cells (e.g. hairs, glandular/oil/resin cells, oil/resin canals), which localized in one or several plant organs (Fig. 1.4).

1.3 Organoleptic, Physical Characteristics and Chemical Compositions

In general, most of EOs are colourless, lucid and mobile liquids at room temperature. However, the entire colour spectrum of EOs actually ranges from yellow to dark brown with all the intermediate scale of colours, excluding the chamomile (Roman) EOs which appears distinctive blue-violet colour due to the presence of chamazulene generated during the steam distillation. These colours are selectively used in perfumes as toners. Moreover, solid such as crystals (e.g. stearoptenes) are found in EOs of rose, chamomile and some *Eucalyptus* species. The typical odour of EOs is dependant of the organs, the species and the origin of plants. Unlike vegetable oils, EOs are volatile oils with a high refractive index and optimal rotation as the result of many asymmetrical compounds. The relative density of EOs is commonly lower than that of water while several exceptions exist. EOs are usually recognized hydrophobic but they are largely soluble in fats, alcohols and most of organic solvents. Besides, they have sensitivity of being oxidized to form resinous products through polymerization.

Fig. 1.4 Localization of secretory structures in common aromatic plants

EOs are comprised of compounds with diverse chemical structures, which are produced in all aromatic plants and trees by photosynthesis through two pathways. One is the multiplication of activated isoprene (isopentenyl pyrophosphate C_5) to its even and uneven addition products. The other is the shikimic acid biosynthesis, in which some biomolecule deviations are responsible for aromatic compounds derived from phenylpropane such as eugenol, anethole, cinnamic aldehyde, etc. Therefore, all EOs' compounds can be divided into two main categories, hydrocarbons mainly the mono-, sesqui- and di-terpenes and oxygenated compounds, for instance,

Table 1.1 Different types of familiar compounds in essential oils

Chemical functions	Example of structure	Molecules	Plants
Hydrocarbon		**Limonene** α-pinene **Phellandrene** β-caryophyllene α-camphorene	Orange, citron, **Geranium, star anise,** *Eucalyptus,* **Clove,** **Camphor tree**
Alcohol		**Linalool** Prenol **Menthol** **Farnesol** **Vetiselinelol** **Phytol** α-terpineol	Ylang ylang, **Mint,** **Lavender,** **Cardamom,** **Chamomile,** **Vetiver,** **Iasmine,** *Citrus*
Phenol		**Eugenol** Thymol **Anethole** **Safrole**	Thym, **Clove,** **Star anise,** **Sassafras**
Ether-oxide		**1,8-cineole** Geranyl butyl ether	Rose, *Eucalyptus,*
Aldehyde		**Geranial** Cinnamic aldehyde **Neral**	Cinnamon, *Citrus,* *Pelargonium*
Ketone		**Carvone** α- and β- vetivone **Menthone**	Vetiver, **Caraway,** **Peppermint**
Ester		**Linalyle acetate** Geranyl acetate **Neryl & α-terpinyl acetate**	*Pelargonium,* Lavender, *Citrus*
Acid		**Benzoic acid** Cinnamic acid	Apple, **Cinnamon,**
Other compounds (nitrogenous and sulphur molecules, lactone, fatty acids, …)		**Coumarin** Indole **Dimethyl trisulfide** **Ambrettolide** **Dillapiole**	Gardenia, Jasmine, **Rose,** **Lavender,** **Ambrette,** **Dill seed**

alcohols, oxides, aldehydes, ketones, phenols, acids and esters/lactones. Some familiar molecules in EOs are classified in Table 1.1 on the basis of their chemical functions (Fernandez et al. 2013). According to the profile of gas chromatography coupling with mass spectrometry (GC-MS), it is common to identify several tens or even hundreds of components in the same essential oil. Nevertheless, some EOs contain only a few compounds that determine the aromatic properties of EOs even their content are low.

The high variability of EOs in terms of their composition and yield can be explained by various parameters belonging to two categories. Intrinsic parameters correspond to the species, organs and ripeness of plants, as well as farming methods, harvesting time and environmental interactions (climate, soil, etc.). Extrinsic parameters include extraction, storage and packaging, among which the extraction process (method, treatment time, etc.) may cause the major alternations in EOs' chemical compositions owing to the hydrolysis. Nowadays, all these factors with accurate analytical data help International Organization for Standardization (ISO) to regulate the world EOs' market. Nonetheless, the adulteration of EOs is of frequent occurrence worldwide because of the urging of commercial interest and the lacking of the knowledge. Hence, it is important to explore extensive and convictive knowledge on EOs' functionality and their quantitative and qualitative analyses, and to train special talents in this field as well.

1.4 Conclusions

The knowledge of EOs has been constantly developed and updated in the long steam of history thanks to a growing number of studies in multidisciplinary fields. Although some compositions and mechanisms are not completely understood, the fascinating odour of EOs and their interesting biological activities are of great interest, which have been pushing the progress of the EOs' research in recent years.

A Scanning electron micrograph of untreated rosemary leaf

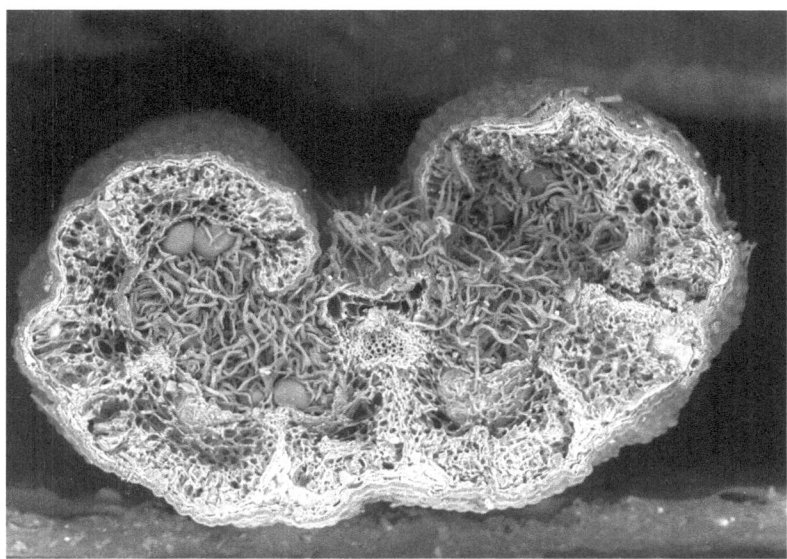

Acknowledgements   Authors thank Prof. Bernard Vidal from University of La Reunion (France) for his valuable comments and discussions about the history of chemistry and natural products.

References

Chemat F, Abert-Vian M, Fernandez X (2013) Microwave-assisted extraction of essential oils and aromas. In: Chemat F (ed) Microwave-assisted extraction for bioactive compounds: theory and practice. Springer, New York, pp 53–66

Fernandez X, Chemat F, Do TKT (2013) Essential oils: virtues and applications. Vuibert, Paris (in French)

French J (1651) The art of distillation. Editions Richard Cotes, London

ISO International Standard 9235 (2013) Aromatic natural raw materials—vocabulary. International Organization for Standardization, Geneva, Switzerland

Macquer M (1756) Elements of theoretical chemistry. Editions J-T. Herissant, Paris (in French)

Smadja J (2009) Essential oils: chemical composition and localization. In: Chemat F (ed) Essential oils and aromas: green extraction and applications. Har Krishan Bhalla & Sons, Dehradun, pp 122–146

Chapter 2
Essential Oils: From Conventional to Green Extraction

Abstract This chapter reviews the development of extraction techniques for essential oils (EOs). The conventional extraction techniques and their intensifications are summarized in terms of their principles, benefits and disadvantages. The green extractions with innovative techniques are also elaborated for future optimization and improvement of traditional EOs' productions.

Keywords Essential oils · Extraction techniques · Optimization · Innovation

2.1 Conventional Extraction

As previously described, EOs are defined as products extracted from natural plants by physical means only such as distillation, cold press and dry distillation. However, the loss of some components and the degradation of some unsaturated compounds by thermal effects or by hydrolysis can be generated by these conventional extraction techniques. These disadvantages have attracted the recent research attention and stimulated the intensification, optimization and improvement of existing and novel "green" extraction techniques. All these techniques are appropriately applied with a careful consideration of plant organs and the quality of final products. Moreover, the analytical composition of EOs extracted from the same plant organ may be quite different with respect to the techniques used. These conventional extraction techniques could typically extract EOs from plants ranging from 0.005 to 10 %, which are influenced by the distillation duration, the temperature, the operating pressure, and most importantly, the type and quality of raw plant materials.

2.1.1 Steam Distillation

Steam distillation is one of ancient and official approved methods for isolation of EOs from plant materials. The plant materials charged in the alembic are subjected to the steam without maceration in water. The injected steam passes through the

© The Author(s) 2014

Y. Li et al., *Essential Oils as Reagents in Green Chemistry*, SpringerBriefs in Green Chemistry for Sustainability, DOI 10.1007/978-3-319-08449-7_2

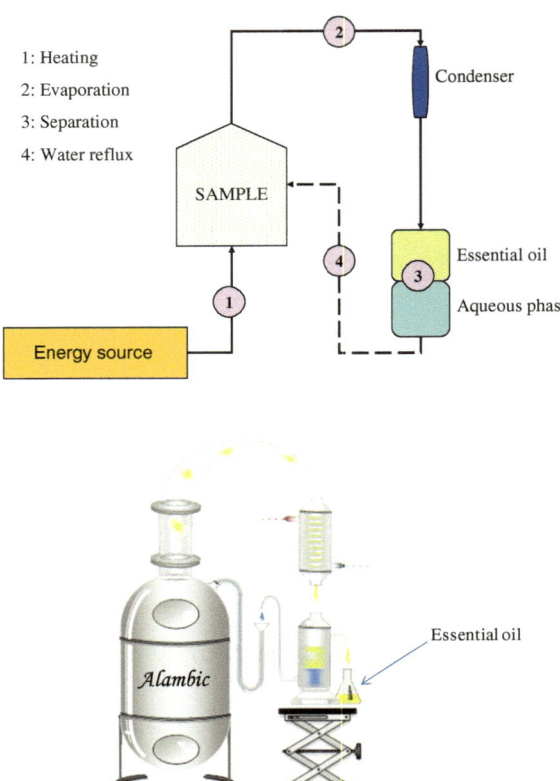

Fig. 2.1 A schematic representation of conventional recovery of essential oils

1: Heating

2: Evaporation

3: Separation

4: Water reflux

plants from the base of the alembic to the top. The vapour laden with essential oils flows through a "swan-neck" column and is then condensed before decantation and collection in a Florentine flask (Fig. 2.1). EOs that are lighter or heavier than water form two immiscible phases and can be easily separated. The principle of this technique is that the combined vapour pressure equals the ambient pressure at about 100 °C so that the volatile components with the boiling points ranging from 150 to 300 °C can be evaporated at a temperature close to that of water. Furthermore, this technique can be also carried out under pressure depending on the EOs' extraction difficulty.

2.1.2 Hydro-diffusion

Unlike steam distillation, the steam injected in this system is from the top of the alembic to the bottom. The vapour mixture with EOs is directly condensed below the plant support through a perforated tray. The way of separating EOs is the same as that in other distillation methods. This method can reduce the steam consumption and the distillation time, meanwhile, a better yield can be obtained in comparison with steam distillation.

2.1.3 Hydro-distillation

Hydro-distillation (HD) is a variant of steam distillation, which is recommended by the French Pharmacopoeia for the extraction of EOs from dried spices and the quality control of EOs in the laboratory. Instead of the steam input, the plant materials in HD are directly immersed in water. This solid-liquid mixture is then heated until boiling under atmospheric pressure in an alembic, where the heat allows the release of odorous molecules in plant cells. These volatile aroma compounds and water form an azeotropic mixture, which can be evaporated together at the same pressure and then condensed and separated in a Florentine flask due to their immiscibility and density difference. Moreover, a cohobation system can recycle the distilled water through a siphon so as to improve the yield and quality of EOs. It is important to mention that the recovered EOs are different from the original essence due to the long treatment duration.

2.1.4 Destructive Distillation

This technique is only applied on birch (*Betula lenta* or *Betula alba*) and cade (*Juniperus oxycedrus*). The toughest parts of these woods (e.g. barks, boughs, roots, etc.) are exposed to dry distillation through a tar after undergoing a destructive process under intense heat. A typical, leathery and empyreumatic oil is then obtained after condensation, decantation and separation.

2.1.5 Cold Expression

This technique is an extraction without heating for EOs of citrus family (Fig. 2.2). The principle of this mechanic process is based on machine squeezing the citrus pericarps at room temperature for the release of EOs, which are washed in cold running water. The essence is then isolated by decantation or centrifugation. Although this method retains a high value of citrus odour, the high consumption of water can affect EOs' quality as the result of the hydrolysis, the dissolution of oxygenated compounds and the transport of microorganism. Several new physical processes appear more popular for the reason of avoiding such deteriorations. The oleaginous cavities on the peel are pressed to burst by two horizontal ribbed rollers (*sfumatrice*) or a slow-moving Archimedian screw coupling to an abrasive shell (*pelatrice*) thus EOs are bent to release. The oil-water emulsion is separated after

Fig. 2.2 Schematic of cold expression

rinse off with a fine spray of water. Besides, the machines which treat citrus peels only after removal of juices and pulps are known as *sfumatrici*, while those which process the whole citrus fruit are called *pelatrici* (Guenther 1948).

2.2 Green Extraction with Innovative Techniques

Since economy, competitiveness, eco-friendly, sustainability, high efficiency and good quality become keywords of the modern industrial production, the development of EOs' extraction techniques has never been interrupted. Strictly speaking, conventional techniques are not the only way for the extraction of EOs. Novel techniques abided by green extraction concept and principles have constantly emerged in recent years for obtaining natural extracts with a similar or better quality to that of official methods while reducing operation units, energy consumption, CO_2 emission and harmful co-extracts in specific cases. The principles of green extraction can be generalized as the discovery and the design of extraction processes which could reduce the energy consumption, allow the use of alternative solvents and renewable/innovatory plant resources so as to eliminate petroleum-based solvents and ensure safe and high quality extracts or products (Chemat 2012).

2.2.1 Turbo Distillation

This technique is developed to reduce energy and water consumption during boiling and cooling in hydro-distillation. The turbo extraction allows a considerable agitation and mixing with a shearing and destructive effect on plant materials so as to shorten distillation time by a factor of 2 or 3. Furthermore, it is an alternative for extraction of EOs from spices or woods, which are relatively difficult to distill.

Besides, an eco-evaporator prototype could be added with aspect of the recovery and the reuse of the transferred energy during condensation for heating water into steam (Chemat 2010).

2.2.2 Ultrasound-Assisted Extraction

With the aim of higher extraction yields and lower energy consumption, ultrasound-assisted extraction has developed to improve the efficiency and reduce the extraction time in the meanwhile. The collapse of cavitation bubbles generated during ultra-sonication gives rise to micro-jets to destroy EOs' glands so as to facilitate the mass transfer and the release of plant EOs. This cavitation effect is strongly dependent to the operating parameters (e.g. ultrasonic frequency and intensity, temperature, treatment time, etc.) which are crucial in an efficient design and operation of sono-reactors. In addition to the yield improvement, the EOs obtained by Ultrasound-Assisted Extraction (UAE) showed less thermal degradation with a high quality and a good flavor (Porto et al. 2009; Asfaw et al. 2005). However, the choice of sonotrode should be careful as the result of the metallic contamination which may accelerate oxidation and subsequently reduce EOs' stability (Pingret et al. 2013). This technique has already proved its potency to scale up, which shows 44 % of increment on extraction yield of EOs from Japanese citrus compared to the traditional methods (Mason et al. 2011).

2.2.3 Microwave-Assisted Extraction

Microwave is a non-contact heat source which can achieve a more effective and selective heating. With the help of microwave, distillation can now be completed in minutes instead of hours with various advantages that are in line with the green chemistry and extraction principles. In this method, plant materials are extracted in a microwave reactor with or without organic solvents or water under different conditions depending on the experimental protocol. The first Microwave-Assisted Extraction (MAE) of EOs was proposed as compressed air microwave distillation (CAMD) (Craveiro et al. 1989). Based on the principle of steam distillation, the compressed air is continuously injected into the extractor where vegetable matrices are immersed in water and heated by microwave. The water and EOs are condensed and separated outside the microwave reactor. The CAMD can be completed in just 5 min and there is no difference in quantitative and qualitative results between extracts of CAMD and 90 min conventional extraction using steam distillation. In order to obtain high quality EOs, vacuum microwave hydro-distillation (VMHD) was designed to avoid hydrolysis (Mengal et al. 1993). Fresh plant materials have been exposed to microwave irradiation so as to release the extracts; reducing the pressure to 100–200 mbar enables evaporation of the azeotropic water-oil mixture at a temperature lower than 100 °C. This operation can be repeated in a stepwise way with a constant microwave power, which is contingent on the desired yield.

The VMHD, which is 5–10 times faster than classic HD, showed comparable yield and composition to HD extracts. The EOs have a organoleptic properties very close to the origin natural materials. Moreover, the occurrence of thermal degradation reduces because of the low extraction temperature. Beyond that, in fact, there exist a couple of modern techniques assisted by microwave such as microwave turbo hydro-distillation and simultaneous microwave distillation, which are impressive for short treatment time and less solvent used (Ferhat et al. 2007; Périno-Issartier et al. 2010).

On account of growing concern for the impact of petroleum-based solvents on the environment and the human body, several greener processes without solvent have sprung up in the last decade. Solvent-free microwave extraction (SFME) was developed with considerable success in consistent with the same principles as MAE (Li et al. 2013). Apart from the benefits mentioned before, the SFME simplifies the manipulation and cleaning procedures so as to reduce labor, pollution and handling costs. The SFME apparatus allows the internal heating of the in situ water within plant materials, which distends the plant cells thus leads to the rupture of oleiferous glands.

A cooling system outside the microwave oven allows the continuous condensation of the evaporated water-oil mixture at atmospheric pressure. The excessive water is refluxed to the reactor in order to maintain the appropriate humidity of plant materials. It is interesting to note that the easy-controlled operating parameters need to be optimized for maximization of the yield and final quality. The potential of using SFME at laboratory and industrial scale has been proved on familiar plant materials with a considerable efficiency compared to conventional techniques (Filly et al. 2014). Inspired by SFME, a number of its derivatives have emerged, which offer significant advantages like shorter extraction time, higher efficiency, cleaner feature, similar or better sensory property under optimized conditions (Michel et al. 2011; Sahraoui et al. 2008, 2011; Wang et al. 2006; Farhat et al. 2011). In 2008, a novel, green technique namely microwave hydro-diffusion and gravity (MHG) has been originally designed (Fig. 2.3). This technique is a microwave-induced hydro-diffusion of plant materials at atmospheric pressure, which all extracts including EOs and water drop out of the microwave reactor under gravity into a continuous condensation system through a perforated Pyrex support. It is worth mentioning that the MHG is neither a modified MAE that uses organic solvents, nor an improved HD that are high energy and water consumption, nor a SFME which evaporates the EOs with the in situ water only. In addition, MHG derivants such as vacuum MHG and microwave dry-diffusion and gravity (MDG) has developed later with the consideration of energy saving, purity of end-products and post-treatment of wastewater (Farhat et al. 2010; Zill-e-Huma et al. 2011).

2.2.4 Instantaneous Controlled Pressure Drop Technology

The DIC process is a direct extraction-separation technique, which is not like the molecular diffusion in conventional techniques. It allows volatile compounds to be removed by both evaporation for a short time at high temperature (180 °C) and high

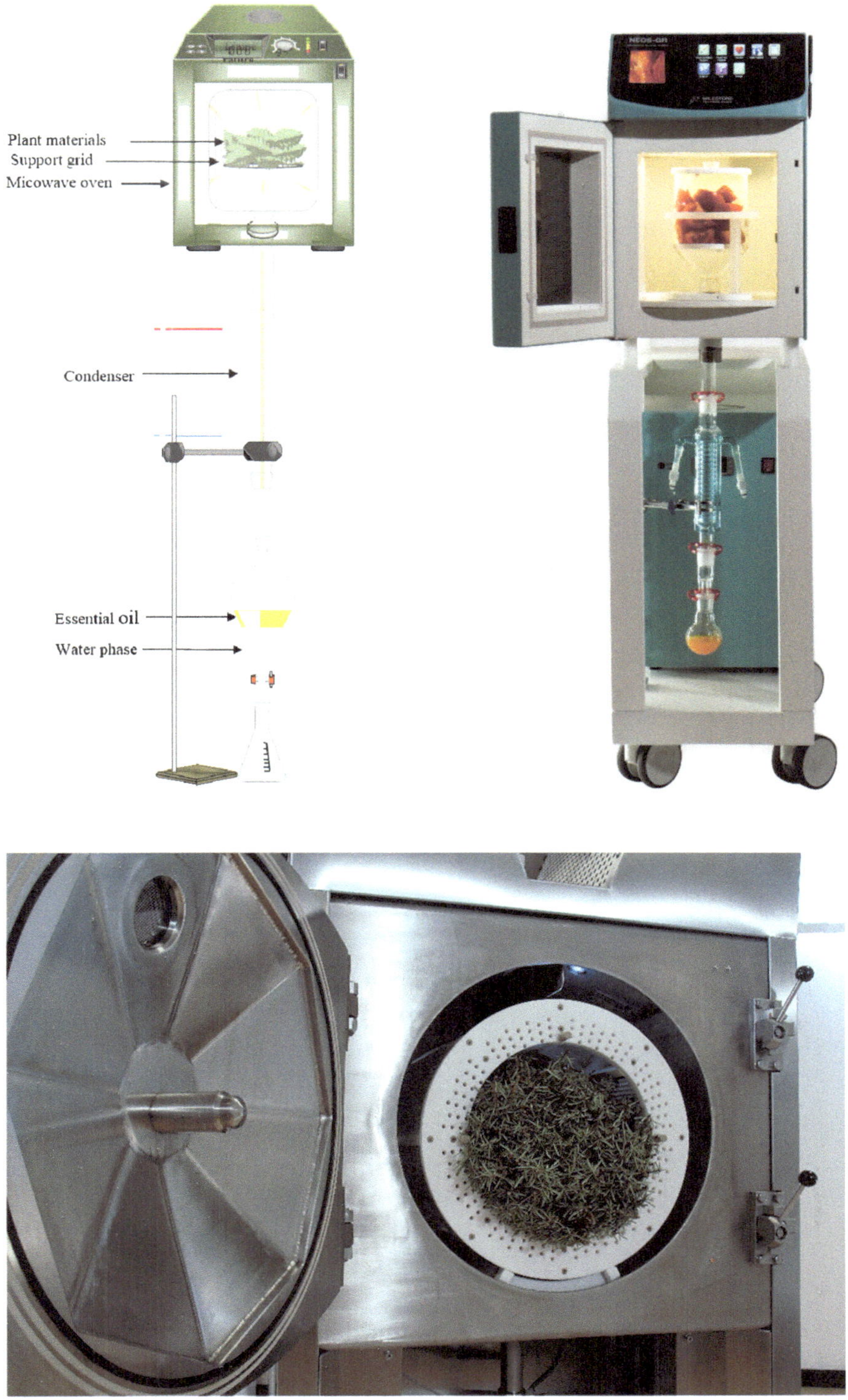

Plant materials
Support grid
Micowave oven

Condenser

Essential oil
Water phase

Fig. 2.3 Solvent free microwave extraction at laboratorial and industrial scale

Table 2.1 Innovative techniques for extraction of essential oils

Name	Brief introduction	Advantages (A) and drawbacks (D)	Main influencing parameters	References
Simultaneous distillation extraction (SDE)	Either HD or steam distillation is combined with solvent extraction, which is frequently used for the isolation of volatile compounds from EOs bearing plants. Solvent used should be insoluble in water and of high purity. SDE has been modified into several variants with the consideration of efficiency, scale and quality of end-products	A: less solvents, elimination of excessive thermal degradation and dilution of extract with water D: artefact production, loss of hydrophilic compounds	Treatment time Solvent Oxygen	Jayatilaka et al. (1995) Blanch et al. (1996) Chaintreau (2001) Altun and Goren (2007) Teixeira et al. (2007)
Pulsed electric field assisted extraction (PEF)	This technique applies short pulses at high voltage in order to create electro compression, which causes plant cells to be ripped open and perforated. The treatment chamber in PEF consists of at least two electrodes with an insulating region in between, where the treatment of plant materials happens	A: preserved fresh character, low heating impact and energy consumption D: only for pumpable materials, restricted by viscosity and particle size of products, high cost	Flow rate Pulse frequency Electric field strength Preheating	Jeyamkondan et al. (1999) Barbosa-Canovas et al. (2000) Fincan et al. (2004)
Supercritical fluid extraction	The plant material is placed in an extractor with the flow of supercritical CO_2. In the supercritical state (above 74 bar and 31 °C), CO_2 is characterized as lipophilic solvent with the high diffusivity, which gives itself a good capacity for diffusion, and a high density ranging from gas-like to liquid-like endows the capacity of transport and major extraction. The fluids carrying extracts pass through the gas phase. The extracts are then separated and collected in a separator	A: inexpensive CO_2, nontoxic, high diffusion, rapidity, selectivity and no denaturation of sensitive molecules D: expensive equipment investment, high energy consumption for pressure and temperature establishment	Treatment time Pressure Flow rate of CO_2	Mira et al. (1996) Reverchon (1997) Caredda et al. (2002) Marongiu et al. (2003) Donelian et al. (2009)

(continued)

Table 2.1 (continued)

Name	Brief introduction	Advantages (A) and drawbacks (D)	Main influencing parameters	References
Subcritical water extraction	The hot water is used at temperatures between boiling (100 °C) and critical point (374.1 °C) of water. Water is maintained in its liquid form under the effect of high pressure. Under these conditions, the polarity of water decreases, which allows the extraction of medium polar and nonpolar molecules without using organic solvents	A: clean, low cost, simple, safe, rapidity, adjustable water polarity, high ratio of oxygenated compounds	Temperature Pressure Water flow rate Solid particle size	Jiménez-Carmona et al. (1999) Ayala and Luque de Castro (2001) Smith (2002) Eikani et al. (2007) Giray et al. (2008)
		D: expensive equipment investment, high energy consumption, thermal degradation		

pressure (10 bar) and auto-vaporization from alveolated plant structures resulting from multi-cycle instantaneous pressure drop (Rezzoug et al. 2005; Besombes et al. 2010). This solvent-free process presents a significant improvement whether in efficiency or in energy consumption and a very short heating time in each DIC cycle eliminate the thermal degradation. Moreover, the DIC obtained the same or even higher yield of EOs with a higher quality than conventional methods regarding to their more oxygenated compounds and lower sesquiterpene hydrocarbons. In addition, heating time and cycle number in particular, have an influence on the extraction efficiency of DIC for all aromatic herbs and spices (Allaf et al. 2013a, b).

2.2.5 Other Emerging Green Extraction Techniques

With the exception of above-described techniques, there are other emerging techniques for EOs extraction which are well established in the early time of the innovation. Table 2.1 summarizes these techniques in terms of their fundamentals, influencing parameters, advantages and draw-backs. It is hard to ignore that all these techniques have been successfully applied at an industrial scale.

2.3 Conclusions

An overview of extraction technique has been presented here for obtaining EOs, which covers a range from conventional to up-to-date methods. The new techniques have been proved to obtain extracts with higher quality in a shorter time compared to traditional techniques. Nevertheless, from a regulatory point of view, these so-called EOs of innovative techniques are not listed in norms due to the restrictive definition of EOs which is only based on the conventional extraction methods. As the consequence of this, the amendment or reestablishment of industry standards in a broader sense becomes more important than ever.

A scanning electron micrograph of untreated lavender

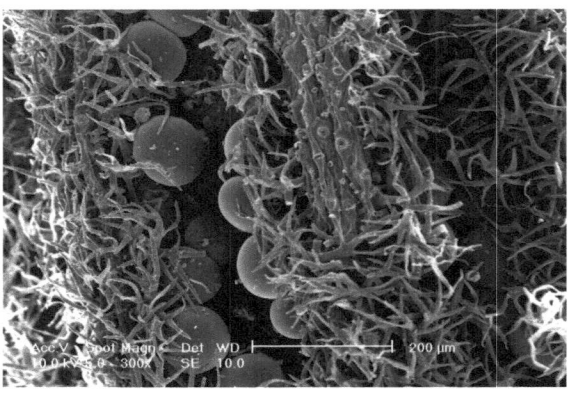

References

Allaf T, Tomao V, Besombes C, Chemat F (2013a) Thermal and mechanical intensification of essential oil extraction from orange peel via instant autovaporization. Chem Eng Process 72:24–30

Allaf T, Tomao V, Ruiz K, Chemat F (2013b) Instant controlled pressure drop technology and ultrasound assisted extraction for sequential extraction of essential oil and antioxidants. Ultrason Sonochem 20:239–246

Altun M, Goren AC (2007) Essential oil composition of *Satureja cuneifolia* by simultaneous distillation-extraction and thermal desorption GC-MS techniques. J Essent Oil Bearing Plants 10:139–144

Asfaw N, Licence P, Novitskii AA, Poliakoff M (2005) Green chemistry in Ethiopia: the cleaner extraction of essential oils from *Artemisia afra*: a comparison of clean technology with conventional methodology. Green Chem 7:352–356

Ayala SR, Luque de Castro MD (2001) Continuous subcritical water extraction as a useful tool for isolation of edible essential oils. Food Chem 52:2335–2338

Barbosa-Canovas GV, Pierson MD, Zhang QH, Schaffner DW (2000) Pulsed electric fields. J Food Sci 65:65–79

Besombes C, Berka-Zougali B, Allaf K (2010) Instant controlled pressure drop extraction of *lavandin* essential oils: fundamentals and experimental studies. J Chromatogr A 1217:6807–6815

Blanch GP, Reglero G, Herraiz M (1996) Rapid extraction of wine aroma compounds using a new simultaneous distillation-solvent extraction device. Food Chem 56:439–444

Caredda A, Marongiu B, Porcedda S, Soro C (2002) Supercritical carbon dioxide extraction and characterization of *Laurus nobilis* essential oil. J Agric Food Chem 50:1492–1496

Chaintreau A (2001) Simultaneous distillation-extraction: from birth to maturity—review. Flavours Fragr J 16:136–148

Chemat F (2010) Techniques for oil extraction. In: *Sawamura M Citrus* essential oils: flavor and fragrance. Wiley, New Jersey pp 9–28

Chemat F (2012) Green extraction of natural products: concept and principles. Int J Mol Sci 13:8615–8627

Craveiro AA, Matos FJA, Alencar JW, Plumel MM (1989) Microwave oven extraction of an essential oil. Flavour Fragr J 4:43–44

De Gioannis B, Marongiu B, Porcedda S (2001) Extraction and isolation of *Salvia desoleana* and *Metha spicata* subsp. *insularis* essential oils by supercritical CO_2. Flavour Fragr J 16:384–388

Donelian A, Carlson LHC, Lopes TJ, Machado RAF (2009) Comparison of extraction of patchouli (*Pogostemon cablin*) essential oil with supercritical CO_2 and by steam distillation. J Supercrit Fluid 48:15–20

Eikani MH, Golmohammad F, Rowshanzamir S (2007) Subcritical water extraction of essential oils from coriander seeds (*Coriandrum sativum L.*). J Food Eng 80:735–740

Farhat A, Fabiano-Tixier AS, El Maataoui M, Maingonnat JF, Romdhane M, Chemat F (2011) Microwave steam diffusion for extraction of essential oil from orange peel: kinetic data, extract's global yield and mechanism. Food Chem 125:255–261

Farhat A, Fabiano-Tixier AS, Visinoni F, Romdhane M, Chemat F (2010) A surprising method for green extraction of essential oil from dry spices: microwave dry-diffusion and gravity. J Chromatogr A 1217:7345–7350

Ferhat M, Tigrine-Kordjani N, Chemat S, Meklati BY, Chemat F (2007) Rapid extraction of volatile compounds using a new simultaneous microwave distillation solvent extraction. Chromatographia 65:217–222

Filly A, Fernandez X, Minuti M et al (2014) Solvent-free microwave extraction of essential oil from aromatic herbs: from laboratory to pilot and industrial scale. Food Chem 150:193–198

Fincan M, De Vito F, Dejmek P (2004) Pulsed electric field treatment for solid-liquid extraction of red beetroot pigment. J Food Eng 64:381–388

Guenther E (1948) The essential oils. Lancaster Press, New York

Giray ES, Kirici S, Kaya DA et al (2008) Comparing the effect of sub-critical water extraction with conventional extraction methods on chemical composition of *Lavandula stoechas*. Talanta 74:930–935

Jayatilaka A, Poole SK, Poole CF, Chichila TMP (1995) Simultaneous micro steam distillation/solvent extraction for the isolation of *semivolatile* flavour compounds from cinnamon and their separation by series coupled-column gas chromatography. Anal Chim Acta 302:147–162

Jeyamkondan S, Jayas DS, Holley RA (1999) Pulsed electric field processing of food: a review. J Food Protect 9:975–1096

Jiménez-Carmona MM, Ubera JL, Luque de Castro MD (1999) Comparison of continuous subcritical water extraction and hydro distillation of *marjoram* essential oil. J Chromatogr A 855:625–632

Li Y, Fabiano-Tixier AS, Abert-Vian M, Chemat F (2013) Solvent-free microwave extraction of bioactive compounds provides a tool for green analytical chemistry. Trends Anal Chem 47:1–11

Marongiu B, Porcedda S, Caredda A et al (2003) Extraction of *Juniperus oxycedrus* ssp. *oxycedrus* essential oil by supercritical carbon dioxide: influence of some process parameters and biological activity. Flavour Fragr J 18:390–397

Mason T, Chemat F, Vinatoru M (2011) The extraction of natural products using ultrasound and microwaves. Curr Org Chem 15:237–247

Mengal P, Behn Dm Bellido M, Monpon B (1993) VMHD: extraction of essential oil by microwave. Parfums Cosmet Aromes 114:66–67 (in French)

Michel T, Destandau E, Elfakir C (2011) Evaluation of a simple and promising method for extraction of antioxidants from sea buckthorn (*Hippophae rhamnoides L.*) berries: pressurised solvent-free microwave assisted extraction. Food Chem 126:1380–1386

Mira B, Blasco M, Subirats S (1996) Supercritical fluid extraction and fractionation of essential oils from orange peel. J Supercrit Fluid 9:238–243

Périno-Issartier S, Abert-Vian M, Petitcolas E, Chemat F (2010) Microwave turbo hydrodistillation for rapid extraction of the essential oil from *Schinus terebinthifolius* Raddi Berries. Chromatographia 72:347–350

Pingret D, Fabiano-Tixier AS, Chemat F (2013) Degradation during application of ultrasound in food processing: a review. Food Control 31:593–606

Porto C, Decorti D, Kikic I (2009) Flavour compounds of *Lavandula angustifolia L.* to use in food manufacturing: comparison of three different extraction methods. Food Chem 112:1072–1078

Reverchon E (1997) Supercritical fluid extraction and fractionation of essential oils and related products. J Supercrit Fluid 10:1–37

Rezzoug SA, Boutekedjiret C, Allaf K (2005) Optimization of operating conditions of rosemary essential oil extraction by a fast controlled pressure drop process using response surface methodology. J Food Eng 71:9–17

Sahraoui N, Abert-Vian M, Bornard I, Boutekdjiret C, Chemat F (2008) Improved microwave steam distillation apparatus for isolation of essential oils: comparison with conventional steam distillation. J Chromatogr A 1210:229–233

Sahraoui N, Abert-Vian M, Elmaataoui M, Boutekdjiret C, Chemat F (2011) Valorization of citrus by-products using microwave steam distillation. Innov Food Sci Emerg Technol 12:163–170

Smith RM (2002) Extraction with superheated water. J Chromatogr A 975:31–46

Teixeira S, Mendes A, Alves A, Santos L (2007) Simultaneous distillation-extraction of high-value volatile compounds from *Cistus Ladanifer L.* Anal Chim Acta 584:439–446

Wang Z, Ding L, Li T et al (2006) Improved solvent-free microwave extraction of essential oil from dried *Cuminum cyminum L.* and *Zanthoxylum bungeanum* Maxim. J Chromatogr A 1102:11–17

Zill-e-Huma H, Abert-Vian M, Elmaataoui M, Chemat F (2011) A novel idea in food extraction field: study of vacuum microwave hydro diffusion technique for by-products extraction. J Food Eng 105: 351–360

Chapter 3
Essential Oils as Antioxidants

Abstract This chapter shows a complete picture of current knowledge on the antioxidant properties of essential oils (EOs). It presents the mechanism of degradation and the influencing factors of EOs' antioxidant activity. Furthermore, various analytical methods used for the evaluation of EOs' antioxidant activities are presented. In addition, application of EOs as antioxidants in food and cosmetic field are also included.

Keywords Essential oils · Antioxidants · Analytical methods · Applications

Lipids are very vulnerable to oxidation, enzymatic or microbial autolysis. The oxidation of lipids, which result in degradation on sensory quality, nutritional or functional value of products during manufacture and storage, can be facilitated by several parameters such as the presence of oxygen free radicals, sunlight, a high temperature, or even the trace of transition metals (Mau et al. 2004). In order to limit the loss of quality and safety, manufacturers have applied some physical practices such as cold storage, vacuum cooking, packaging under an inert atmosphere, or the addition of synthetic antioxidants such as butylated hydroxyanisole (BHA), butylated hydroxytolune (BHT), *tert*-butylhydroquinone (THBQ) and propyl gallate (PG), which is a relatively easy and economical way to inhibit the oxidative deterioration of lipids. However, these synthetic antioxidants are suspected to have mutagenic, carcinogenic and teratogenic effects during a long-term use (Chavéron 1999). With the growing concern about the safety use of these preservatives and the increasing consumer demand for green natural products without additives, emerging significant studies have led industries to consider the incorporation of non-chemical substances in their food or cosmetic preparation.

EOs are considered as potential resources of natural bioactive molecules, which have been numerously investigated for their antioxidant properties (Bakkali et al. 2008). The EOs extracted from aromatic plants are generally used as food flavours or additives through a simple addition way (Karpinska et al. 2001). They are mostly classified as generally recognized as safe (GRAS), which have also been approved as food additives by Food and Drug Administration in USA (Hulin et al. 1998). EOs

Y. Li et al., *Essential Oils as Reagents in Green Chemistry*, SpringerBriefs in Green Chemistry for Sustainability, DOI 10.1007/978-3-319-08449-7_3

are liposoluble, non-toxic antioxidants that can effectively retard lipid peroxidation and minimize rancidity at a low dose without affecting the quality of the products. Moreover, an ideal antioxidant must also be stable during various technological processes (Pokorny et al. 2001). Therefore, in order to be able to use EOs to their full potential as antioxidants with appropriate doses for product conservation, it is essential to understand the mechanism of lipid oxidation in food products and relevant influencing parameters, as well as the assessment of their antioxidant efficacy.

3.1 Mechanism of Degradation

Degradation is generally undesirable since it can change the organoleptic, nutritional or functional characteristics of foods and cosmetics. There are numerous factors influencing or initiating lipid oxidation, which are classified with intrinsic [e.g. unsaturation of lipid fatty acids (number and position), the presence of pro-oxidants (metal ions, enzymes, etc.) and natural antioxidants] and external factors (e.g. temperature, light, partial pressure of oxygen, water activity, and conditions of storage and processing).

The lipid oxidation can result from the following three pathways: autoxidation caused by temperature, metal ions and free radicals; photo-oxidation initiated by light in the presence of photosensitizer; and enzymatic oxidation initiated by lipoxygenase (Fig. 3.1). The autoxidation is a series of radical reactions involving three stages. The first initiation stage produces a lipid free radical known as alkyl radical (R$^{\cdot}$) by abstraction of a hydrogen atom from the fatty acid (RH) in the presence of an initiator. This slow reaction is promoted with an increasing temperature, which is easily be induced by ionizing radiation, chemical generators, enzymatic or chemical systems that can produce reactive oxygen species or metallic traces. Subsequently, the triplet oxygen ($^3O_2$) is quickly fixed to these free radicals to form unstable peroxyl free radicals (ROO$^{\cdot}$) that can further transform to lipid hydroperoxides (ROOH) through hydrogen abstraction from other new fatty acids. All free radicals produced in this propagation stage combine each other to form non-radical species (ROOR, RR, O_2) in the termination stage. The photosensitizers (Sens) in photo-oxidation absorb light energy for transformation of their excited triplet state (Sens^3), which are involved in the lipid oxidation with their two types. The first-type photosensitizer such as riboflavin acts as a free radical initiator. In its excited state, the L$^{\circ}$ formation occurs by the abstraction of a hydrogen atom or an electron from a fatty acid, which is capable of reacting with triplet oxygen ($^3O_2$) to get ROOH. The second-type photosensitizers, such as chlorophyll and erythrosine, react with $^3O_2$ in their excited state for transfer their energy to obtain a singlet oxygen ($^1O_2$), which is highly electrophilic and bond directly up to an unsaturated fatty acid (RH) thereby a ROOH is formed. There are two main enzymes (lipoxygenase and cyclooxygenase) in the enzymatic oxidation. The lipoxygenase helps the insertion of an oxygen molecule to an unsaturated fatty acid through a stereospecific reaction, which results in the ROOH formation. However, it acts specifically on non-esterified fatty acids and its activity is often

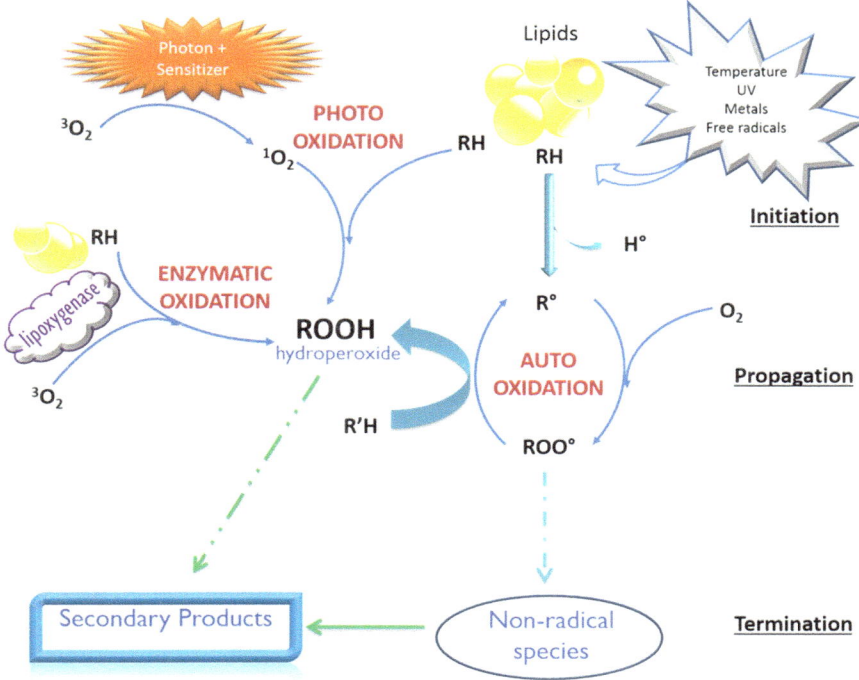

Fig. 3.1 Lipid oxidation pathways: autoxidation, photo-oxidation and enzymatic oxidation

related to that of lipases and phospholipases. Cyclooxygenase is a lipoxygenase which can incorporate two oxygen molecules into a fatty acid in order to form a specific hydroperoxide. The enzymatic oxidation occurs even at low temperatures (0–4 °C). The enzyme activity is very low during the frozen storage whereas it recovers and rises once thawing process begins.

3.2 Main Assessment of Antioxidant Activity

An antioxidant is a compound that can efficiently protect a given target against oxidation while being used at a very low antioxidant/target molar ratio, i.e., lower than 1 %. In foods, the most important targets are polyunsaturated lipids (RH), where lipid autoxidation is typically initiated at the water-lipid interface of food emulsions by low-valence transition metal traces (Fe^{II}, Cu^{I}). These prooxidant species can be autoxidized to their high-valence counterparts (Fe^{III}, Cu^{II}) with the subsequent formation of superoxide ($O_2^{\cdot-}$) and hydrogen peroxide, or trigger the homolytic cleavage of hydroperoxides (hydrogen peroxide, traces of lipid hydroperoxides) with the subsequent formation of highly oxidizing oxyl radicals ($HO^{\cdot}$, $RO^{\cdot}$).

An antioxidant that inhibits lipid peroxidation can act according to several mechanisms, in which the most important are shown as follows,

- The fast scavenging of lipid peroxyl radicals (ROO˙) that propagate the peroxidation chain (inhibition of propagation). Such antioxidants are called chain-breaking antioxidants. A common example is α-tocopherol (vitamin E), which can transfer its phenolic H-atom to LOO˙ and is then simultaneously converted into a resonance-stabilized radical. Unlike the LOO˙, this radical is unable to react with a second RH molecule to propagate the chain.
- Acting as peroxidation initiators (e.g. $O_2^{˙-}$, HO˙) in the scavenging of hydrophilic oxidizing species formed in the aqueous phase (inhibition of initiation). Hydrophilic antioxidants such as ascorbate and polyphenols can behave as initiation inhibitors.
- The regeneration of a potent chain-breaking antioxidant by reduction of the corresponding radical. Ascorbate and several polyphenols are proved to be able to regenerate α-tocopherol at a water-lipid interface.
- The formation of inert transition metal complexes that are unable to carry out the radical-generating processes mentioned above. Citric acid and polyphenols having a catechol group (1,2-dihydroxybenzene) may eventually act according to this mechanism.

From this simplified overview, it is clear that the antioxidant activity is governed by several factors such as the reducing activity of the antioxidant (ability to quickly deliver H-atoms and/or electrons), its location in biphasic water-lipid systems (roughly predicted from the hydrophilic-lipophilic balance), and its metal binding capacity. Consequently, antioxidant tests may give highly contrasted results depending on whether they involve mono- or bi-phasic systems and they make use of metal ions or organic species (e.g., diazo compounds) for radical generation. Hence, the results of a single assay only give a reductive view of the antioxidant properties of EOs and must be interpreted with caution. Moreover, EOs are complex mixtures of compounds with different HLB and reducing capacity, which makes data interpretation even more complicated and offers more opportunities for scattered results depending on the antioxidant tests selected.

Although a variety of physiochemical methods exist nowadays for assessing the antioxidant activity of natural extracts (Chemat et al. 2007; Decker et al. 2005; Somogyi et al. 2007), the antioxidant capacities of EOs differ from one method to another due to EOs' diverse nature and the complexity of oxidation processes, which makes it impossible to compare between methods and certain standardizations afterwards. The frequently-used way to have a relatively accurate value of the EOs' antioxidant capacity is combination of several results of antioxidant test for interpretations. The antioxidant activity can be assessed either by direct measuring of formed products (e.g. hydroperoxides) or by indirect measuring of EOs' ability to scavenge free radicals with an intermediate probe.

Two main approaches can be applied to determine in vitro antioxidant activities. Firstly, the inhibition of lipid autoxidation in different systems (oil, solutions of lipids in organic solvents, oil-in-water emulsions, micelles, liposomes, lipoproteins, etc.). These methods are aimed at modelling oxidation processes in foods or living systems. Therefore, they are close to real situations and, being generally based on biphasic systems, offer the opportunity to test both hydrophilic and lipophilic antioxidants. Their main drawbacks are time-consuming experimental procedures and complexity in data interpretation. Indeed, lipid autoxidation is typically slow and yields a complex distribution of products beyond the primarily formed lipid hydroperoxides such as lipid alcohols and carbonyl compounds, cleavage products, volatile compounds, etc. Measurable and reproducible autoxidation rates are more readily achieved when large concentrations of initiator (diazo compounds, metal ions) are used. Overall, the data may depend on the type of initiator used, the lipid autoxidation products monitored and the analytical technique selected. The protective action of EOs has been evaluated in two model systems by measuring the formation of primary (conjugated dienes) and secondary (TBARS) lipid autoxidation products. Methods that can be used to measure inhibition of lipid oxidation are peroxide value determination (AOCS Cd 8–53 1997), p-anisidine value (ISO 6885 2006), Conjugated dienes method (AOCS Cd 7–58 2009; Klein 1970), thiobarbituric acid reactive substance (TBARS) method (AOCS Cd 19–90 2009) and β-carotene bleaching test (Taga et al. 1984).

Secondly, the ability of antioxidants to scavenge free radicals species can be easily generated or are eventually stable enough to be commercially available and handled as common chemicals. Although these methods do not afford clear biological significance and must be interpreted with caution, they are typically much simpler and easier to implement than methods based on lipid autoxidation. They can offer first efficient approaches for screening a large number of samples and are widely popular in the agro-food industry. Different methods like Trolox Equivalent Antioxidant Capacity (TEAC) method (Miller et al. 1993), Ferric Reducing Antioxidant Potential (FRAP) method (Oyaizu 1986; Iris et al. 1999), DPPH (2,2-diphenyl-1-picrylhydrazyl) method, Oxygen Radical Absorbance Assay (ORAC), Total Radical Trapping Antioxidant Parameter (TRAP) method (Wayner et al. 1985), Electron Paramagnetic Resonance (EPR), can be used to achieve this radical scavenging test. These methods are based on the ability of EOs to scavenge free radicals, such as the superoxide radical anion ($O_2^{\cdot-}$), the hydroxyl radical (HO·) and the stable coloured radicals $ABTS^{\cdot+}$, DDPH·.

3.3   Application of Essential Oils as Antioxidants

The preservation of food and cosmetic products in a green and efficient way has been a hot-spot research for industry. It is essential to prevent from microbial contamination and lipid oxidation in order to guarantee a product with a sufficient shelf life. Nowadays, more and more EOs have been qualified as natural antioxidants and proposed as potential substitutes to synthetic antioxidants in practical applications. A number of studies have already been conducted to prove that some essential oils from natural plants can not only play a key role in limiting the lipid oxidation of meat and other fatty foods in particular (Estévez et al. 2007; Botsoglou et al. 2002; Botsoglou et al. 2003; Goulas and Kontamoinas 2007), but also contribute to the development of a pleasant odour and favourable taste for consumers. However, this inhibition effect is dependent on the food type and the storage condition as well. In cosmetic formulations containing oils rich in unsaturated fatty acids, the formulators are forced to add antioxidants and required to replace synthetic preservatives for the sake of healthy and safe products without rancidity. The effectiveness of antioxidants in hydrophobic systems depends on their solubility, stability and volatility. Antioxidants are better to be liposoluble so that they can react with free radicals in lipid oxidation. Moreover, they should also be thermally stable during production processes. In addition, they should not be too volatile in risk of being lost during the manufacturing process of the final products, which hampers the use of essential oils as antioxidants (Branen and Davidson 1996).

3.4 Conclusions

As the result of that lipid oxidation involving sensory impairments is of great concern to the food and cosmetic industry. Antioxidants, which have important health implications, have become an essential part of preservation strategies for final products. More and more manufacturers have recently realized the negative health effects of synthetic antioxidant and they are urged to search and use natural antioxidants, especially plant extracts. Several EOs have been developed and proposed to use as natural antioxidant in recent years due to their interesting antioxidant activity. The antioxidant activity of EOs is better to interpret with several test results since their complex mixtures including several tens of components with distinct hydrogen-donating capacities. Besides, the choice of EOs as antioxidants in industry is also based on several parameters, e.g. solubility, persistent antioxidant ability, sensitivity to pH, the influence on any discoloration of the products, the production of the unpleasant odours or off-flavours, as well as their availability, multiple function and cost.

A scanning electron micrograph of untreated dried orange peels

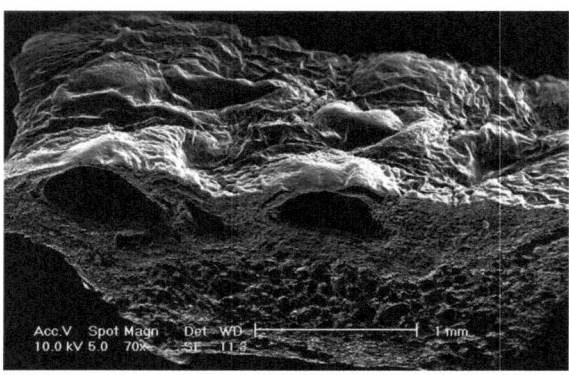

References

American Oil Chemists Society Cd 19–90 (2009) 2-Thiobarbituric acid value direct method. Official Methods and Recommended Practices of the American Oil Chemists' Society, Champaign

American Oil Chemists Society Cd 7–58 (2009) Polyunsaturated acid, ultraviolet spectrophotometric method. Official Methods and Recommended Practices of the American Oil Chemists' Society, Champaign

American Oil Chemists Society Cd 8–53 (1997) Peroxide value-acetic acid-chloroform method, 4th edn. Official Methods and Recommended Practices of the American Oil Chemists' Society, Champaign

Bakkali F, Averbeck S, Averbeck D, Idaomar M (2008) Biological effects of essential oils. A review. Food Chem Toxicol 46:446–475

Botsoglou NA, Christaki E, Fletouris DJ et al (2002) The effect of dietary oregano essential oil on lipid oxidation in raw and cooked chicken during refrigerated storage. Meat Sci 62:259–265

Botsoglou NA, Grigoropoulou SH, Botsoglou E et al (2003) The effects of dietary oregano essential oil and α-tochopheryl acetate on lipid oxidation in raw and cooked turkey during refrigerated storage. Meat Sci 265:1193–1200

Branen AL, Davidson P (1996) Use of antioxidants in self-preserving cosmetic and drug formulation. In: Kabara JJ, Orth DS (eds) Preservative-free and self-preservating cosmetics and drugs: principles and practices. Marcel Dekker, New york, pp 159–179

Chavéron H (1999) Introduction to nutritional toxicology. Lavoisier, TEC & DOC, Paris, p 98 (in French)

Chemat F, Abert-Vian M, Dangles O (2007) Essential oils as antioxidants. Int J Essent Oil Ther 1:4–15

Decker EA, Warner K, Richards MP (2005) Measuring antioxidant effectiveness in food. J Agric Food Chem 53:4303–4310

Estévez M, Ramırez R, Ventanas S, Cava R (2007) Sage and rosemary essential oils versus BHT for the inhibition of lipid oxidative reactions in liver pâté. LWT-Food Sci Technol 40:58–65

Goulas AE, Kontamoinas MG (2007) Combined effect of light salting, modified atmosphere packaging and oregano essential oil on the shelf-life of sea bream (*Sparus aurata*): biochemical and sensory attributes. Food Chem 100:287–296

Hulin V, Mathot AG, Mafart P, Dufosse L (1998) Antimicrobial properties of essential oils and flavour compounds. Sciences des aliments 18:563–582

Iris F, Benzi F, Strain S (1999) Ferric reducing antioxidant assay. Method Enzymol 292:15–27

ISO International Standard 6885 (2006) Animal and vegetable fats and oils-determination of anisidine value. International Organization for Standardization, Geneva

Karpinska M, Borowski J, Danowska-Oziewicz M (2001) The use of natural antioxidants in ready-to-serve food. Food Chem 72:5–9

Klein RA (1970) The detection of oxidation in liposome preparations. Biochim Biophys Acta 210:483–486

Mau JL, Huang PN, Huang SI, Chen CC (2004) Antioxidant properties of methanolic extracts from two kinds of *Antrodia camphorate mycella*. Food Chem 86:25–31

Miller NJ, Rice-Evans C, Davies MJ et al (1993) A novel method for measuring antioxidant capacity and its application to monitoring the antioxidant status in premature neonates. Clin Sci 84:407–412

Oyaizu M (1986) Studies on products of browning reaction prepared from glucosamine. Jpn J Nutr 44:307–315

Pokorny J, Yanishlieva N, Gordon M (2001) Antioxidants in food. Woodhead Publishing Limited and CRC Press, Cambridge

Somogyi A, Rosta K, Pusztai P et al (2007) Antioxidant measurements. Physiol Meas 28:R41–R45

Taga MS, Miller EE, Pratt DE (1984) Chia seeds as a source of natural lipid antioxidants. J Am Oil Chem Soc 61:928–931

Wayner D, Burton G, Ingold K, Locke S (1985) Quantitative measurement of the total peroxyl radical-trapping antioxidant capability of human blood plasma by controlled lipid peroxidation. FEBS lett 187:33–37

Chapter 4
Essential Oils as Antimicrobials

Abstract This chapter presents current knowledge on the antimicrobial activities of essential oils (EOs). It shows the antimicrobial action mechanism of EOs and relevant influencing parameters, thus further elucidates the antimicrobial properties of EOs. Moreover, commonly-used analytical methods for EOs' antimicrobial activities are elaborated. In addition, the application of EOs as natural antimicrobials in various fields has also been introduced.

Keywords Essential oils · Antimicrobial · Analytical methods · Application

EOs and other extracts from aromatic and medicinal plants are empirically known for their antimicrobial properties since ancient times, which have not been scientifically proven until the early of 20th century. The use of EOs has grown over the past four decades and nowadays they have been considered as potential alternatives to antibiotics in chemical preservatives in food and cosmetics and treatment of various infectious diseases. Since the EOs' antibacterial property was firstly investigated by De la Croix in 1881, many other researches on EOs' chemical composition and their antimicrobial activities have been extensively reported (Burt 2004; Chemat 2009; Solorzano-Santos and Miranda-Novales 2011). Generally, the EOs' antimicrobial property is evaluated by two categories depending on the type of microorganisms and bioactive molecules: the inhibitory or bacteriostatic ability for multiplication of microbial cells; and lethal or microbicidal activity for microbial cells. However, it is better to know the mechanism of EOs' antimicrobial action and its influencing factors before further analyses and applications.

4.1 Mechanism of Antimicrobial Action

The physiological role of EOs has not entirely understood yet. The majority of previous works focus more on the detection of EOs' antimicrobial activity, hence, the mechanism of action against microorganisms are poorly studied, most of which are

Y. Li et al., *Essential Oils as Reagents in Green Chemistry*, SpringerBriefs in Green Chemistry for Sustainability, DOI 10.1007/978-3-319-08449-7_4

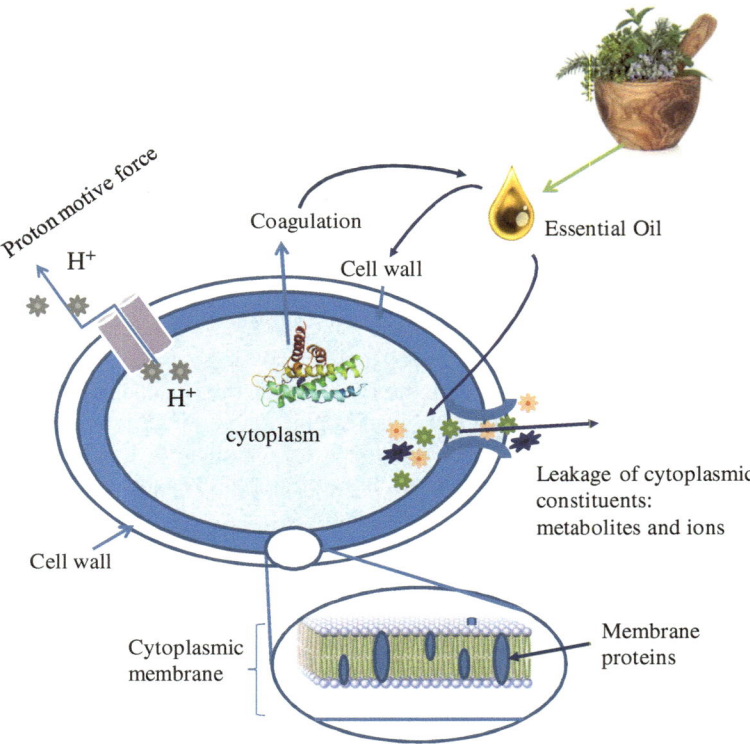

Fig. 4.1 Mechanism for EOs' action in the bacterial cell

assumptions. The mechanism of antimicrobial action seems to have relationship with a great number of complex constituents in EOs instead of just specific bioactive metabolites, which may result in different action modes and difficult identification from molecular point of view (Carson et al. 2002; Burt 2004). In general, EOs' antimicrobial actions are described in three phases (Ultee et al. 2002; Turina et al. 2006). Firstly, EOs spreading on the cell wall of a bacterial enhances the membrane permeability which leads to a subsequent loss of cellular components. The second corresponds to an acidification inside the cell which blocks the production of cellular energy (ATP) due to the ion loss, the collapse of proton pumps and the reduction of membrane potential (Fig. 4.1). Last but not the least is the destruction of genetic materials that results in the death of bacteria. Furthermore, some studies have reported that EOs can also coagulate the cytoplasm and damage lipids, proteins, cell walls and membranes which can lead to the leakage of macromolecules and the lysis afterwards (Gustafson et al. 1998; Cox et al. 2000; Lambert et al. 2001; Di Pasqua et al. 2006; Turgis et al. 2009; Saad et al. 2013).

The effectiveness among different EOs is less obvious than the varied sensitivity of microorganisms to plant extracts. This variability may mainly due to different environmental factors, plant growth, geographic sources, harvesting seasons, genotypes, climates, drying procedures, studied plant organs, extraction and analytical methods of biological activities. All above-mentioned parameters affect the chemical composition and relative concentration of each active component in EOs, which play a significant role to their antimicrobial activities. In addition, synergic and antagonistic interaction between EOs' chemical components should be considered as well due to the fact that minor components may have significant influence on EOs' antimicrobial activities.

4.2   Antimicrobial Activities of Essential Oils

The antimicrobial effects of various plant species have been empirically used for a long time in order to disinfect for increasing the shelf life of foods. Generally, the antimicrobial property of plants is directly related to its fraction of EOs contained, which is variable and complex due to the number and the nature of chemically different molecules inside (Oussalah et al. 2007). The antimicrobial activity is the result of functional groups presenting in the metabolites and their synergies. The most active functional group is phenols, followed by aldehydes, ketones, alcohols, ethers and hydrocarbons (Gao et al. 2005). The antimicrobial activities of various EOs have been studied and described that most of effective EOs belong to the *Lamiaceae* family, including thyme, oregano, rosemary, lavender, mint, etc. (Burt 2004).

4.2.1   Antibacterial Properties

EOs and their major compounds have a broad spectrum of action against of a wide range of bacteria including those are resistance to antibiotics such as methicillin-resistant *Staphylococcus*, vancomycin-resistant *Enterococci* and ciprofloxacin-resistant *Campylobacter* (Fisher and Philips 2009; Varder-Ünlü et al. 2006). Their effectiveness is variable between EOs and bacteria strains. The distinction between Gram-positive and negative bacteria is based on a difference in the cell wall composition. The structure of wall in Gram-positive bacteria is homogenous with a thickness varying from 10 to 80 nm; it is rich in teichoic acid and osmaines but poor in lipids (<2 %). On the contrary, the wall structure in Gram-negative bacteria is more complex with a thickness of about 10 nm and it is rich in lipids (10–22 %) and contains less osmaines (Barton 2005).

EOs have generally been observed to be more active against Gram-positive bacteria than Gram-negative bacteria though there are some exceptions (Oussalah et al. 2007). The Gram-negative bacteria *Aeromonas hydrophila* and *Campylobacter jejuni* have been found to be particularly sensitive to the EOs' action. However, there is no general rule of sensitivity for Gram bacterial due to many controversies existing in previous published works (Burt 2004). It is now well known that the chemical compositions in EOs from a particular plant species may vary depending on the geographical origin, harvesting or extraction methods, which are main causes of variability in the sensitive degree of Gram-positive and negative bacteria to EOs. The compounds with the highest antibacterial efficacy and the broadest spectrum are phenols such as thymol, eugenol and carvacrol (Dorman and Deans 2000; Nevas et al. 2004).

4.2.2 Antifungal Properties

The antifungal activity of EOs has been demonstrated in several studies. Guynot et al. (2005) had studied twenty EOs' effects against the most important moulds regarding the spoilage of bakery products. The results showed that only cinnamon leaf, rosemary, thyme, bay and clove EOs exhibited antifungal activity against all fungus, which suggest a possibility of using EOs as alternatives to synthetic chemicals in preservation of bakery products. The more active constituents (carvacrol and thymol) in oregano essential oil performed highest fungicidal activity so that this essential oil was identified as the most active against various fungi. Origanum EOs have proved fungistatic and fungicidal to a human pathogenic yeast (*Candida albicans*), of which a daily oral administration are highly recommended in the effective prevention and treatment of candidiasis (Manohar et al. 2001). Moreover, carvacrol and eugenol were proposed as therapeutic agents for oral candidiasis because of their potent antifungal properties (Chami et al. 2004). Besides, the effectiveness of 5 antifungal compounds had been evaluated on twelve fungi, which citral and geraniol were the most active, followed by linalool, cineole and menthol (Pattnaik et al. 1997). Concerning the *Penicillium* fungi, the highly effective inhibition of their mycelium growth was found for mugwort essential oil and a complete inhibition of toxin production was observed (Khaddor et al. 2006). Some other studies indicated that EOs from some medicinal plants may become potential candidates to prevent edible products from the growth of toxigenic fungus and subsequent toxin contamination (Razzaghi-Abyaneh et al. 2009).

4.3 Main Assessment of the Antimicrobial Activity

As the result of the diverse methodologies for evaluation of EOs' antimicrobial activity, it should be careful to select appropriate techniques for determining their antimicrobial activity. The insolubility of EOs' constituents in the water and their volatility are the main practical difficulties, which may explain the variety of techniques. Apart from the most applied mass spectrometry-based techniques (GC-FID, GC-MS) which allow an identification, comparison and quantification of EOs' components, conventional in vitro methods in solid or liquid media can characterize the antimicrobial potency and quantify the minimum inhibitory or microbicide concentration, while a novel technique called 'omic' explored a convenient way for advanced assessment of genetic sequence and expression profile, and protein functional content of single species as well.

4.3.1 Micro-atmosphere Method in Vapour Phase

This method highlights the action of volatile EOs' components against germ development on a solid medium in a Petri dish instead of quantifying the real antimicrobial activity of EOs. The microorganism is inoculated on the surface of an agar and a few drops of EOs are deposited on a blotting paper or a small cup placed at the bottom and the centre of the cover. The Petri dish is inversely incubated at optimum temperature (Fig. 4.2). The absence of microbial growth in a translucent area on the agar with a clear contour reveals the antimicrobial action of EOs' volatiles through their evaporation (Lopez et al. 2005; Becerril et al. 2007).

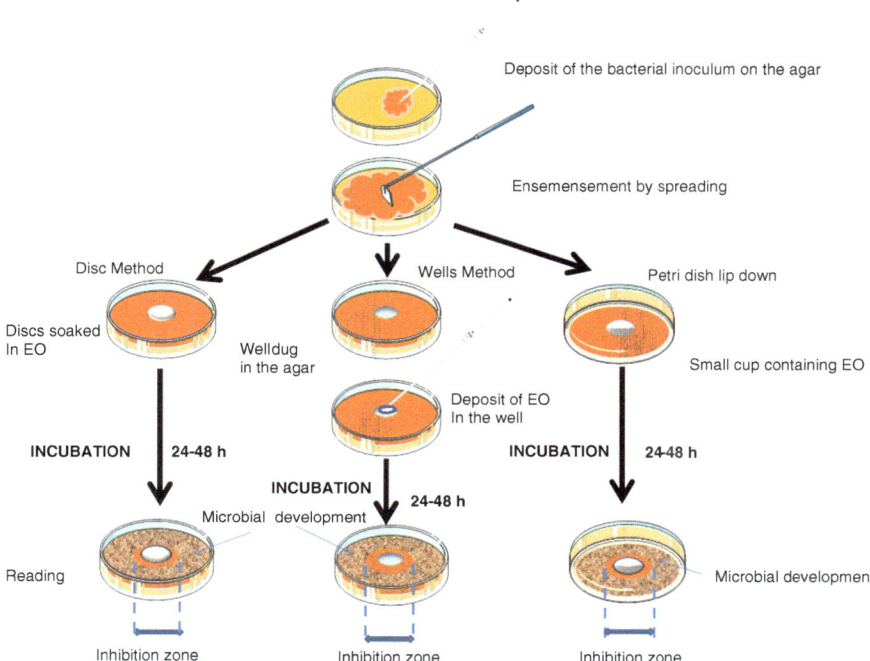

Fig. 4.2 Analysis of antimicrobial activity by micro-atmosphere and agar diffusion

4.3.2 Diffusion Methods on Solid Media

These qualitative methods test susceptibility or resistance of microorganisms by direct contact with EOs. Aromatograms, namely disk diffusion method, is similar to antibiograms for antibiotics test, which can evaluate EOs' antibacterial activities. A blotting paper disk of 6 mm diameter impregnated with diluted EOs of 15 μl is deposited on the surface of an agar medium previously inoculated with the studied microbial suspension. All prepared Petri dishes were then incubated under optimal conditions of microorganism cultivation. The germs grow as visible colonies during incubation so that a clear halo around the disk indicates the inhibition of microbial growth, of which the diameter of the inhibition halo depends on the sensitivity to EOs and are measured in mm. Two controls were performed under the same operating conditions: a negative control with the solvent used to solubilize EOs and an antibiotic disk is used as a positive control (Fig. 4.2). The well method is used for screening of large amounts of EOs and/or microbial isolates. The only difference is that a sterile well of 6 mm diameter filled with EOs using a micropipette is applied instead of blotting paper in disk method (Fig. 4.2). These two methods are usually used for the pre-selection of the EOs' antimicrobial activity because the inhibition diameter is not a direct measure of the EOs' activity but a qualitative indication of the sensitivity or resistance of germs. A classification of EOs chemotyped concerning their spectrum of antimicrobial activity may be determined by the importance of the inhibition halo. However, these techniques have limitations due to various factors in terms of EOs' constituents, inoculum density, culture thickness, deposited volume of EOs, solvent nature for solubilization of EOs and microbial strains.

4.3.3 Dilution Method (Broth and Agar)

The purpose of this method is to determine the lowest concentration of antimicrobials that inhibit the growth of the tested bacteria. This method is carried out in a series of broth or agar media containing decreasing concentrations of antimicrobial agents with a standardized microbial suspension (generally equivalent to 10^8 bacteria/ml). Taking broth culture as example, the mother solution in the first tube or Petri dish contains 400 μl of test EOs with known concentration and 4.6 ml of sterile broth culture. A control is separately prepared in broth by adding 2.5 ml of solvent alone used for solubilisation. The serial dilution is then performed by adding 2.5 ml of the mother solution to 2.5 ml of broth contained in the second test tube. The rest of dilutions can be done in the same manner so as to obtain a range of decreasing concentration. 15 μl of inoculum is then introduced into each tube and all tubes are incubated at the optimal temperature for 24–48 h, which can help prove to have sensitivity to EOs by the previous methods.

The minimum inhibitory concentration (MIC) expressed as μl/ml or mg/l has been measured by optical density measurement at 600 nm for determining the end-point of bacterial growth.

4.3.4 Determination of Minimum Inhibitory and Bactericidal Concentration

The inhibition diameter and minimum inhibitory concentration (MIC) are often characterized for a strain in term of its resistance or sensitivity to antibiotics in medical bacteriology. The MIC is the lowest antibiotic concentration for complete inhibition of bacterial growth up to 24 or 48 h incubation, which is extensively used to quantify antibacterial activity of EOs (Canilac and Mourey 2001; Delaquis et al. 2002). Its various definitions in different studies are obstacles for result comparison. The MIC is frequently not bactericidal because the inoculum cells will develop when the antibacterial agent disappears.

The minimum bactericidal concentration (MBC) is the concentration of required inhibitor for a complete bactericidal action. It is defined as the lowest concentration at which no growth is observed after subculturing into a fresh broth (Onawunmi 1989). It is the concentration that can generate ≥99.9 % mortality of the initial microbial cells. It is determined in liquid or solid medium by evaluating the survivors after the removal of the antibiotics (Rota et al. 2008). The MBC test succeeds the MIC test directly, in which the cultures without bacteria growth as well as control tube are inoculated again into new liquid or solid media in Petri dishes, followed by a 24–48 h incubation under optimal temperature for the target microorganism. The minimum concentration of EOs required for no bacterial development after incubation is considered the MBC (Fig. 4.3).

4.3.5 'Omic' Techniques

The 'omic' techniques in terms of genomics, transcriptomics and proteomics have been developed to follow the antimicrobial activities of EOs and their components against single pathogen in particular (Demissie et al. 2011; Sertel et al. 2011; Efferth and Koch 2011). These novel techniques coupled with meta-genomics, transcriptomics and proteomics help to understand how microbial communities respond to changes in their environment so that they could become a prerequisite for better understanding of the EOs' action mechanism by identification of their targets and the disrupted molecular functions and pathways (Gillbert and Hughes 2011; Mitra et al. 2011). The use of 'omic' techniques will accelerate the search for active compounds among a wide number of analyses and facilitate the selection of specific ligands for single cellular and molecular processes (Fig. 4.4).

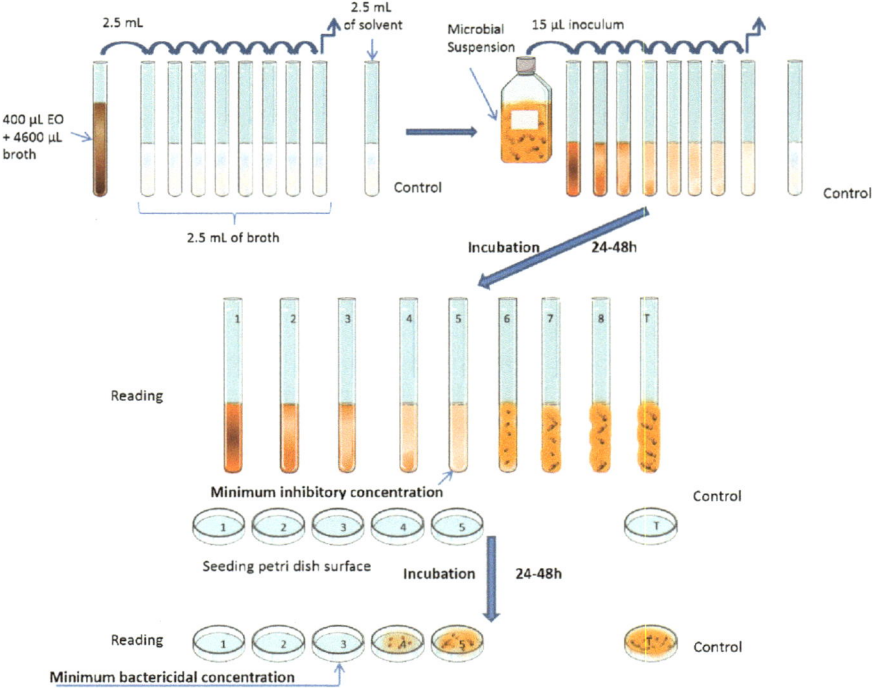

Fig. 4.3 Determination of MIC on liquid media and MBC in solid media

4.4 Application of EOs as Antimicrobials

Although the use of plant EOs has been primarily studied in the medical field, the natural properties of EOs and their effectiveness to new applications have been explored and developed over thirty years, especially in food, cosmetic and public health fields.

EOs from various herbs and spices have been widely acknowledged to be food flavourings and preservatives because of their aromatic and antimicrobial constituents inside. They have great potential to be natural preservatives that can be added in almost all foods due to their high efficiency at very low concentrations (Vrinda Menon and Garg 2001; Rooler and Seedhar 2002). Moreover, the incorporation of EOs directly into foods or spraying of EOs on the food surface helps to control the microbial flora for prevention of food oxidation (Oussalah et al. 2006). Besides, EOs have been used in the food packaging for inhibition of spoilage moulds and food pathogens (Nielsen and Rios 2000; Becerril et al. 2007). The application of edible polymers (biofilm, capsule, emulsion, coating) containing EOs can significantly reduce the microorganism growth during storage. It is essential to notice the influence of environmental conditions, interactions between EOs and food constituents, physical structures and compositions of food matrix on EOs' efficiency

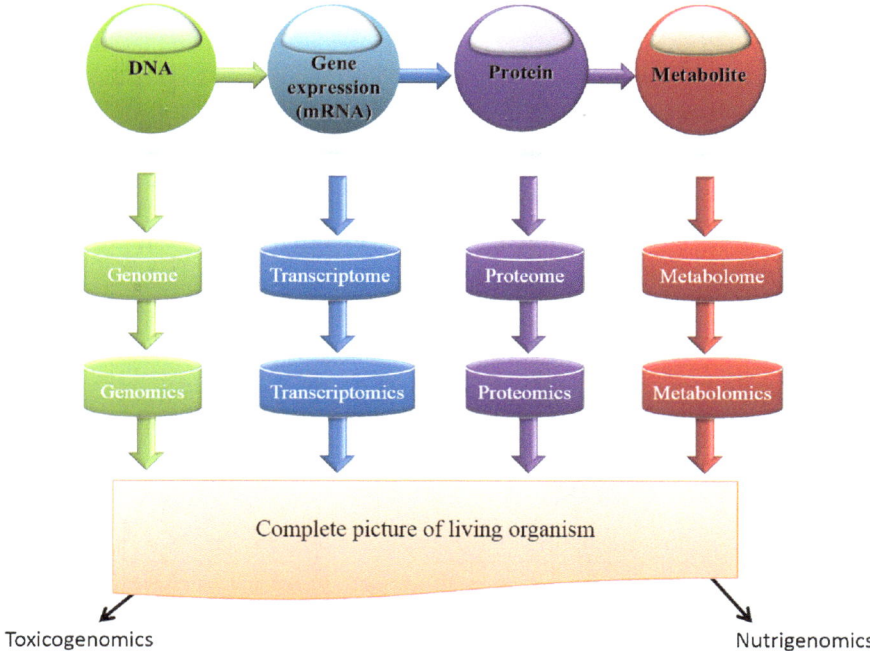

Fig. 4.4 Interacting of 'omic' technologies

against pathogens in foods, as well as the organoleptic properties of the food products with addition of EOs (Gill et al. 2002; Chouliara et al. 2008).

The scientific phytotherapy uses experimental techniques to emphasize the relationship between the chemical structures of EOs' bioactive molecules and their functional properties. As mentioned previously, the value of EOs is attributed to all their constituents instead of only major components. The main therapeutic antimicrobial activities of plant extracts have been summarized as follows (Pisseri et al. 2008). EOs of *Melaleuca alternifolia, Thymus* spp., *Satureja* spp., *Citrus bergamia, Origanum vulgaris, Illicium verum, Ocimum basilicum, Matricaria recutita, Salvia officinalis, Satureja montana, Origanum majorana* for their in vitro antibacterial properties and EOs of *Melaleuca alternifolia, Thymus vulgaris, Origanum vulgaris, Citrus lemon, Ocimum basilicum, riganum majorana* for their in vivo and in vitro antifungal properties. The EOs' antimicrobial properties have inspired various studies of air disinfection in hospitals, which a microbiological clean ventilation system is implemented using EOs for air circulation. EOs of *Satureia montana* L., *Thymus vulgaris* CT *thymol, Origanum vulgaris* and *Cinnamomum verum* have proved their bactericidal capacities against *Staphylococcus aureus* and *Pseudomonas aeruginosa*, which are two main pathogens for typical nosocomial infections (Pibiri 2005). Besides, the effectiveness of EOs has also been embodied against plant pathogenic bacteria in a sustainable agricultural strategy (Ryan 2002).

EOs have also been widely used in dermatology for treatment of skin infections in particular. For instance, the EOs of lavender are often incorporated into soaps and other cosmetic products for treatment of dermatoses and skin maintaining. Since the synthetic preservatives authorized in annex V of the Cosmetics Directive have been gradually avoided or banned for the sake of health and less risk, EOs as natural preservatives have attracted more attention to be alternatives in cosmetics due to their antimicrobial properties. Nonetheless, the concentration of EOs used is generally higher than the recommend percentage in cosmetics, which may generate unfit odors for the final product. In addition, EOs at these concentrations may induce the allergic skin reaction. Therefore, the use of EOs is strictly regulated by the scientific committee on consumer products (SCCP) for constituents as potential allergens in EOs which should be labeled on the product. The 7th amendment of Cosmetics Directive has regulated that compulsory labels should be required if EOs present more than 0.001 % in washing products (soaps, shampoos, etc.) and more than 0.01 % in non-washing products (creams, etc.).

4.5 Conclusions

EOs and their constituents generally display efficient antimicrobial properties, which have been used as preservatives against various microbial diseases. More studies should be carried out on synergism and antagonism of constituents in EOs and foods before using these substances. Moreover, the degree of toxicity should be first investigated when EOs are employed for conservation and therapeutic aims, especially in fields of food and cosmetic.

A scanning electron micrograph of untreated thyme

Acknowledgment Authors thank Prof. Amina Hellal from Ecole Nationale Polytechnique Algeria for her valuable comments and discussion about anti-microbial activity of EO's.

References

Barton LL (2005) Structural and functional relationships in prokaryotes. Springer, USA

Becerril RR, Gómez L, Goñi P et al (2007) Combination of analytical and microbiological techniques to study the antimicrobial activity of a new active food packaging containing cinnamon or oregano against *E. coli* and *S. aureus*. Anal Bioanal Chem 388:1003–1011

Burt S (2004) Essential oil: their antibacterial properties and potential applications in foods. A Rev Int J Food Microbiol 94:223–253

Canillac N, Mourey A (2001) Antibacterial activity of the essential oil of *Picea excelsa* on *Listeria, Staphylococcus aureus* and coliform bacteria. Food Microbiol 18:261–268

Carson CF, Mee BJ, Riley TV (2002) Mechanism of Action of *Melaleuca alternifolia* (Tea Tree) Oil on *Staphylococcus aureus* determined by time-kill, lysis, leakage, and salt tolerance assays and electron microscopy. Antimicrob Agents Ch 46:1914–1920

Chami N, Chami F, Bennis S et al (2004) Antifungal treatment with carvacrol and eugenol of oral candidiasis in immunosuppressed rats. Braz J Infect Dis 8:217–226

Chemat F (2009) Essential oils and aromas: green extractions and applications. H.K. Bhalla & Sons, India

Chouliara E, Badeka A, Savvaidis I, Kontominas MG (2008) Combined effect of irradiation and modified atmosphere packaging on shelf-life extension of chicken breast meat: microbiological, chemical and sensory changes. Eur Food Res Technol 226:877–888

Cox SD, Mann CM, Markham JL et al (2000) The mode of antimicrobial action of essential oil of *Melaleuca alternifola* (tea tree oil). J Appl Microbiol 88:170–175

Delaquis PJ, Stanich K, Girard B, Mazza G (2002) Antimicrobial activity of individual and mixed fractions of dill, cilantro, coriander and eucalyptus essential oils. Int J Food Microbiol 74:101–109

Demissie ZA, Sarker LS, Mahmoud SS (2011) Cloning and functional characterization of β-phellandrene synthase from *Lavandula angustifolia*. Planta 233:685–696

Di Pasqua R, Hoskins N, Betts G, Mauriello G (2006) Changes in membrane fatty acids composition of microbial cells induced by addiction of thymol, carvacrol, limonene, cinnamaldehyde, and eugenol in the growing media. J Agric Food Chem 54:2745–2749

Dorman HJD, Deans SG (2000) Antimicrobial agents from plants: antibacterial activity of plant volatile oils. J Appl Microbiol 88:308–316

Efferth T, Koch E (2011) Complex interactions between phytochemicals. The multi-target therapeutic concept of phytotherapy. Curr Drug Targets 12:122–132

Fisher K, Phillips K (2009) In vitro inhibition of vancomycin-susceptible and vancomycin-resistant *Enterococcus faecium* and *E. faecalis* in the presence of citrus essential oils. Br J Biomed Sci 66:180–185

Gao Y, Jin YJ, Li HD, Chen HJ (2005) Volatile organic compounds and their roles in bacteriostasis in five conifer species. J Integr Plant Biol 47:499–507

Gilbert JA, Hughes M (2011) Gene expression profiling: metatranscriptomics. Methods Mol Biol 733:195–205

Gill AO, Delaquis P, Russo P et al (2002) Evaluation of antilisterial action of cilantro oil on vacuum packed ham. Int. J. FoodMicrobiol 73:83–92

Gustafson JE, Liew YC, Chew S et al (1998) Effects of tea tree oil on *Escherichia coli*. Lett Appl Microbiol 26:194–198

Guynot ME, Marĺn S, SetÚ L, Sanchis V, Ramos AJ (2005) Screening for antifungal activity of some essential oils against common spoilage fungi of bakery products. Food Sci Technol Int 11:25–32

Khaddor M, Lamarti A, Tantaoui-Elaraki A et al (2006) Antifungal activity of three essential oils on growth and toxigenesis of *Penicillium aurantiogriseum* and *Penicillium viridicatum*. J Essent Oil Res 18:586–589

Lambert RJW, Skandamis PN, Coote P, Nychas GJE (2001) A study of the minimum inhibitory concentration and mode of action of oregano essential oil, thymol and carvacrol. J Appl Microbiol 91:453–462

Lopéz P, Sanchez C, Batlle R, Nerin C (2005) Solid and vaporphase antimicrobial activities of six essential oils: susceptibility of selected foodborne bacterial and fungal strains. J Agric Food Chem 53:6939–6946

Manohar V, Cass I, Gray J et al (2001) Antifungal activities of origanum oil against *Candida albicans*. Mol Cell Biochem 228:111–117

Mitra S, Rupek P, Richter D et al (2011) Functional analysis of metagenomes and metatranscriptomes using SEED and KEGG. BMC Bioinform 12(1):S21

Nevas M, Korhonen AR, Lindstrom M et al (2004) Antibacterial efficiency of Finnish spice essential oils against pathogenic and spoilage bacteria. J Food Protect 67:199–202

Nielsen PV, Rios R (2000) Inhibition of fungal growth on bread by volatile components from spices and herbs, and the possible application in active packaging, with special emphasis on mustard essential oil. Int J Food Microbiol 60:219–229

Onawunmi GO (1989) Evaluation of the antimicrobial activity of citral. Lett Appl Microbiol 9:105–108

Oussalah M, Caillet S, Saucier L, Lacroix M (2006) Inhibitory effects of selected plant essential oils on *Pseudomonas putida* growth, a bacterial spoilage meat. Meat Sci 73:236–244

Oussalah M, Caillet S, Saucier L, Lacroix M (2007) Inhibitory effects of selected plant essential oils on the growth of four pathogenic bacteria: *E. coli* O157:H7, *Salmonella typhimirium*, *Staphylococcus aureus* and *Listeria monocytogenes*. Food Control 18:414–420

Pattnaik S, Subramanyam VR, Bapaji M, Kole CR (1997) Antibacterial and antifungal activity of aromatic constituents of essential oils. Microbios 89:39–46

Pibiri MC (2005) Assainissement microbiologique de l'air et des systèmes de ventilation au moyen d'huiles essentielles. Thesis, Ecole polytechnique fédérale de Lausanne, pp 159–179

Pisseri F, Bertoli A, Pistelli L (2008) Essential oils in medicine: principles of therapy. Parassitologia 50:89–91

Razzaghi-Abyaneh M, Shams-Ghahfarokhi M, Mohammad-Bagher R et al (2009) Chemical composition and antiaflatoxigenic activity of *Carum carvi* L., *Thymus vulgaris* and *Citrus aurantifolia* essential oils. Food Control 20:1018–1024

Roller S, Seedhar P (2002) Carvacrol and cinnamic acid inhibit microbial growth in fresh-cut melon and kiwifruit at 4 °C and 8 °C. Lett Appl Microbiol 35:390–394

Rota MC, Herrera A, Martínez RM et al (2008) Antimicrobial activity and chemical composition of *Thymus vulgaris*, *Thymus zygis* and *Thymus hyemalis* essential oils. Food Control 19:681–687

Ryan MF (2002) Insect Chemoreception: fundamental and applied. Kluwer Academic Publishers, Dordrecht, pp 193–222

Saad NY, Muller CD, Lobstein A (2013) Major bioactivities and mechanism of action of essential oils and their components. Flavour Fragr J 28:269–279

Sertel S, Eichhorn T, Plinkert PK, Efferth T (2011) Chemical Composition and antiproliferative activity of essential oil from the leaves of a medicinal herb, Levisticum officinale, against UMSCC1 head and neck squamous carcinoma cells. Anticancer Res 31:185–191

Solorzano-Santos F, Miranda-Novales MG (2011) Essential oils from aromatic herbs as antimicrobial agents. Curr Opin Biotechnol 23:1–6

Turgis M, Han J, Caillet S, Lacroix M (2009) Antimicrobial activity of mustard essential oil against *Escherichia coli* O157:H7 and *Salmonella typhi*. Food Control 20:1073–1079

Turina AV, Nolan MV, Zygadlo JA, Perillo MA (2006) Natural terpenes: self-assembly and membrane partitioning. Biophys Chem 122:101–113

Ultee A, Bennink MHJ, Moezelaar R (2002) The phenolic hydroxyl group of carvacrol is essential for action against the food-borne pathogen *Bacillus cereus*. Appl Environ Microb 68:1561–1568

Vardar-Ünlü G, Ünlü M, Dönmez E, Vural N (2006) Chemical composition and *in vitro* antimicrobial activity of the essential oil of *Origanum minutiflorum*. J Sci Food Agri 87:255–259

Vrinda Menon K, Garg SR (2001) Inhibitory effect of clove oil on Listeria monocytogenes in meat and cheese. Food Microbiol 18:647–650

Chapter 5
Essential Oils as Insecticides

Abstract This chapter presents principal techniques for determining the insecticidal activities of essential oils (EOs) after a brief introduction of the mechanism for EOs' actions. Furthermore, applications of EOs as insecticides against various insects have been summarized with valuable remarks.

Keywords Essential oils · Insecticidal activity · Mortality · Analytical methods · Application

The aromatic plants and their extracts have been empirically used as phytosanitary agents in difference ancient cultures through the ages. The first plant-derived compounds have been identified and isolated in 19th century even though their insecticidal use could trace back to 17th century. Their insecticidal mechanism and market use depending on their toxicity have been scientifically investigated worldwide (Boutekedjiret 2009). However, the natural insecticides have been gradually replaced by the effective synthetic ones due to the insufficient supply of plant raw materials during the 20th century. As the massive and sometimes irrational use of synthetic insecticides generated in the recent past years, people are becoming increasing aware of their adverse effects on human health and environment apart from their effectiveness in the crop protection. Most of these synthetic insecticides are non-degradable and in fact they will accumulate and are persistent in the environment or human bodies through food chains, which often cause chronic diseases and other severe physiological disorders (Saiyed et al. 2003; Lemaire et al. 2004; Fisk et al. 2001; Baldi et al. 2003; Oliva et al. 2001). Since these chemicals have been more stringently restricted by several European regulations, it is essential to explore novel, safe and eco-friendly substitutes with a considerable insecticidal effectiveness.

Y. Li et al., *Essential Oils as Reagents in Green Chemistry*, SpringerBriefs in Green Chemistry for Sustainability, DOI 10.1007/978-3-319-08449-7_5

5.1 Insecticidal Mechanism of EOs

Various studies of natural insecticides showed the secondary metabolites (EOs, terpenoids, polyphenols, steroids and alkaloids) synthesized by plants are responsible for phyto-protective activity against plant pathogens and pests, among which EOs exhibited their significant insecticidal, nematicidal, acaricidal and larvicidal properties. EOs not only act as poisons or neurotropes on the nervous system of insects, but also can intervene in the cellular breathing either by inhibiting cellular oxidation through transfer interruption in the respiratory chain or by asphyxiation through the formation of an impermeable film insulating insects from the air. Moreover, they may have an inhibitory power to enzyme activity and insect growth with respect to adults, larvae and eggs as well.

5.2 Main Assessment of Insecticidal Activity

The choice of analytical method has a significant influence on the determination of EOs' insecticidal activities, which are mainly divided into micro-atmosphere, direct contact and ingestion methods (Fig. 5.1).

Micro-atmosphere method involves a disk of EOs-impregnated filter paper depositing at the centre of Petri dish cover without a direct contact with insects. An evaporation of EOs' volatile compounds then occurs in a hermetical container with insects inside. As the name of the direct contact method implies, EOs can be directly deposited or sprayed on insects in a hermetical container or insects are directly placed in a Petri dish with an EOs-impregnated filter paper in a sealed container. In ingestion method, the insects are put in contact with grains impregnated by EOs. Regardless of the technique used, the reading is relevant to the number of dead insects after a definite exposure time to EOs.

As a result of the fact that the natural mortality exists in any population, the number of dead insects in a treated population is not the real mortality caused by EOs. Therefore, the mortality must be corrected using Shneider-Orelli or Abbot formula with due consideration of the natural death

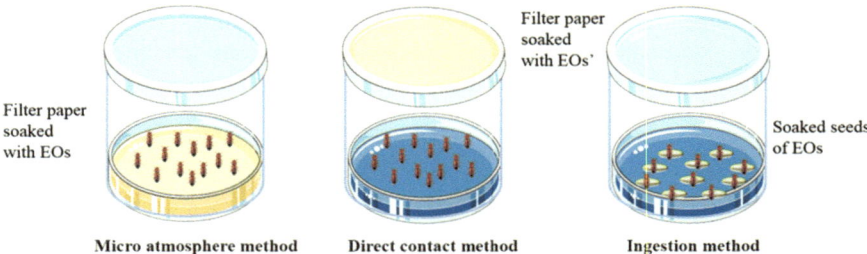

Fig. 5.1 Main assessment of EOs' insecticidal activity

$$Mc(\%) = \frac{M - Mt}{100 - Mt} \times 100$$

where Mc is the corrected mortality, M is the mortality in the treated population and Mt is the mortality in the control population.

The determination of the mortality caused by EOs is not enough to evaluate EOs' insecticidal activity. The further quantification of their effectiveness can be determined in two ways. The first is to determine the mortality according to the increasing doses of EOs (x_n, n = 1, 2, 3,..., n), which leads to establish an experimental curve representing the function Y = f(x) for LD_{50} and LC_{50} calculation. These two values are defined as the dose or lethal concentration that causes 50 % of mortality. The second is to determine the mortality with a constant dose or concentration according to the increasing times (t_n, n = 1, 2, 3...n), which leads to establish an experimental curve representing the function Y = f(t) for LT_{50} calculation at each time. This value is the lethal time corresponding to a 50 % of mortality. LD_{90} or LD_{95} can also use to quantify the necessary amounts of EOs for killing 90 or 95 % population, which is useful for insecticide characterization so as to get rid of pests at utmost.

5.3 Application of Essential Oils as Insecticides

For the sake of discovery of effective and eco-friendly alternatives to synthetic insecticides, the insecticidal properties of various EOs have been widely studied (Table 5.1). The EOs have presented its high effectiveness on not only the adults of various vermin such as beetles, mosquitoes, flies, louses, mites, etc., but also on larvae and eggs of several insects (Prajapati et al. 2005; Daizy et al. 2008). They exhibited herbicidal effects on some noxious weeds as well (Singh et al. 2005). It is worthwhile to mention that all constituents in EOs are responsible for the insecticidal properties of EOs even though the major compounds seem to be more related.

Table 5.1 The insecticidal activity of essential oils

Source of EOs	Family	Insect (adults, eggs and larvae)	Remarks	References
Chamaecyparis obtusa	*Cupressaceae*	*Callosobruchus chinensis* L. and *Sitophilus oryzae* L.	The contact and fumigant activity of EOs varied with its constituents and dose rather than increasing exposure time beyond 1 or 2 days, which the toxicity of these components was significantly decreased at 0.18 mg/cm^2	Park et al. (2003)
Cinnamomum cassia, Cocholeria aroracia and *Brassica juncea*	*Lauraceae, Brassicaceae*	*Sitophilus oryzae (SO)* and *Callosobruchus chinensis*	All EOs caused 100 % mortality within 1 day after treatment with a filter paper diffusion method at 3.5 mg/cm^2 against both species. All EOs were more effective in closed containers than in open ones in a fumigation test with *SO*	Kim et al. (2003a, b)
Cymbopogon schoenanthus	*Poaceae*	*Callosobruchus maculatus*	Piperitone as the main constituent in EOs performed more effective than the crude EOs against adults, eggs and larvae of *C. maculatus*, but less toxic than the crude EOs to individuals developing inside the seeds	Ketoh et al. (2006)
Eucalyptus nicholii, E. codonocarpa, E. blakelyi, Callistemon sieberi, Melaleuca fulgens and *M. armillaris*	*Mirtaceae*	*Sitophilus oryzae (SO), Tribolium castaneum (TC)* and *Rhyzopertha dominica (RD)*	The LD$_{50}$ and LD$_{95}$ against *SO* adults were between 19.0–30.6 and 43.6–56.0 μl/l air, respectively. These six EOs were about twice as toxic to *TC* and *RO* at the LD$_{95}$. 1,8-cineole in EOs shown significant fumigant effects	Lee et al. (2004)
Laurus nobilis	*Lauraceae*	*Rhyzopertha dominica* and *Tribolium castaneum*	The repellent and fumigant toxicities of EOs were highly dependent upon insect species and the origin of EOs	Jemâa et al. (2012)
Melaleuca quinquenervia L. and *Ocimum gratissimum* L.	*Myrtaceae* and *Lamiaceae*	*Callosobruchus maculatus* Fab.	Higher mortality was exhibited for *M. quinquenervia* essential oil at lower concentration (6.66 μl/l) but similar mortality was obtined for both EOs at high concentration (33.3 μl/l). Better insecticidal activity in female *C. Maculatus*	Seri-Kouassi et al. (2004)

(continued)

Table 5.1 (continued)

Source of EOs	Family	Insect (adults, eggs and larvae)	Remarks	References
Ocimum basilicum L. and *Ocimum gratissimum* L.	Lamiaceae	*Callosobruchus maculatus*	80 % mortality was found for *O. basilicum* in a 12 h fumigation using pure EOs at 25 μl/vial, whereas 70 % for *O. gratissimum* and 0 % for control. All EOs had significant effects on both egg hatch rate and a greater sensitivity with males than females was observed	Kéita et al. (2001)
Ocimum gratissimum L.	Lamiaceae	*Sitophilus oryzae* L., *Tribolium castaneum* Herbst, *Oryzaephilus surinamensis* L., *Rhyzopertha dominica* F. and *Callosobruchus chinensis* L.	The fumigant toxicity and repellence of the EOs and their constituents (β-(Z)-ocimene and eugenol) were significantly influenced by concentration and time after treatment. All insects had percentage repellence (PR) values for the EOs (37.5–100 %) and eugenol (45–100 %) except *C. chinensis* showed a dose-dependent decrease in PR values	Ogendo et al. (2008)
Artemisia sieberi	Asteraceae	*Callosobruchus maculatus* (CM), *Sitophilus oryzae* (SO), and *Tribolium castaneum* (TC)	The mortality of old adult insects increased with concentration and exposure time. 100 % mortality was obtained after 24 h with a concentration of 37 μl/l, which CM ($LC_{50} = 1.45$ μl/l) presented much more sensitive than SO ($LC_{50} = 3.86$ μl/l) and TC ($LC_{50} = 16.76$ μl/l)	Negahbana et al. (2007)

(continued)

Table 5.1 (continued)

Source of EOs	Family	Insect (adults, eggs and larvae)	Remarks	References
Chenopodium ambrosioides	*Chenopodiaceae*	*Callosobruchus chinensis, C. maculatus, Acanthosce -lides obtectus, Sitophilus granarius, S. zeamais* and *Prostephanus truncatus*	80–100 % mortality of the beetles was induced during 24 h exposure with EOs of 0.2 μl/cm² with the exception of *C. maculatus* (20 %) and *S. zeamais* (5 %)	Tapondjou et al. (2002)
Cocholeria aroracia, Brassica juncea and *Cinnamomum cassia*	*Brassicaceae* and *Lauraceae*	*Lasioderma serricorne* F.	All EOs exhibited high toxic to the adult beetles after 1 day treatment at 0.7 mg/cm² in the contact test. However, only EOs of *C. aroracia* and *B. juncea* showed more effective insecticidal activity in closed containers than in open ones during fumigation	Kim et al. (2003a, b)
Flourensia oolepis Blake	*Asteraceae*	*Tribolium castaneum* Herbst *Myzus persicae* Sulzer and *Leptinotarsa decemlineata* Say	EOs showed repellent and toxic to *T. castaneum* adult while the behavioral sensibility was found for other two insects	Garcia et al. (2007)
Lavandula angustifolia, Rosmarinus officinalis, Thymus vulgaris and *Laurus nobilis*	*Lamiaceae* and *Lauraceae*	*Sitophilus oryzae (SO), Rhyzopertha dominica (RD)* and *Tribolium castaneum (TC)*	The insecticidal activities of EOs varied with insect species, components inside and the exposure time: highly effective for 1,8-Cineole, borneol and thymol against *SO* at 0.1 μl/720 ml for 24 h while camphor and linalool achieved 1 00 % mortality against *RD* in the same conditions; no significant effect on *TC* after 24 h exposure even at highest concentration but higher mortality was observed after 7 days	Rozman et al. (2007)

(continued)

Table 5.1 (continued)

Source of EOs	Family	Insect (adults, eggs and larvae)	Remarks	References
Origanum glandulosum Desf.	*Lamiaceae*	*Rhizopertha dominica*	The mortality increased with the EOs' concentration used. For total EOs (2 h extraction) and 5 min extracts, the contact effect was better than that of fumigation, while 2 min extracts and 10 min extracts showed significant effect for fumigation with a dose (≤1.56 %) and for contact test with a dose of (1.56 %)	Khalfi et al. (2008)
Salvia hydrangea	*Lamiaceae*	*Sitophilus granaries* and *Tribolium confusum*	EOs showed 68.3–75.0 % mortality against adults	Kotan et al. (2008)
Thymus numidicus Poiret	*Lamiaceae*	*Rhizopertha dominica*	The mortality of crude EOs are weaker than that of 2.5 min extracted EOs due to the presence of linalool and thymol	Saidj et al. (2008)
Cinnamomum osmophloeum	*Lauraceae*	*Aedes albopictus, Culex quinquefasciatus* and *Armigeres subalbatus*	Cinnamaldehyde type leaf EOs and its effective constituent (*trans*-cinnamaldehyde) had great mosquito larvicidal activity	Cheng et al. (2009a)
Cryptomeria japonica	*Taxodiaceae*	*Aedes aegypti* (AE) a nd *Aedes albopictus* (AA)	EOs from the leaves of 58-year-old plant exhibited the most effective larvicidal against both mosquitos. 3-carene and terpinolene respectively showed the best larvicidal activity against *A. aegypti* and *A. albopictus* among 11 constituents	Cheng et al. (2009b)

(continued)

Table 5.1 (continued)

Source of EOs	Family	Insect (adults, eggs and larvae)	Remarks	References
Elletaria cardamomum	Zingiberaceae	Sitophilus zeamais (SZ) and Tribolium castaneum (TC)	The sensitivity of SZ was similar to that of TC at LD_{50} level but higher than TC at LD_{95} level in contact study, while SZ's sensitivity was twice higher at both LD_{50} and LD_{95} level. For TC, the susceptibility of its larvae to contact toxicity increased with age but less sensitive than its adults to fumigation. The survival rate of its larvae, egg hatching and adult emergence was suppressed	Huang et al. (2000)
Eucalyptus saligna	Myrtaceae	Sitophilus oryzae L.	EOs showed the most potent toxicity ($LD_{50} = 28.9$ μl/l air)	Lee et al. (2001)
Eucalyptus saligna and Cupressus sempervirens	Myrtaceae and Cupressaceae	Sitophilus zeamais (SZ) and Tribolium confusum	Eucalyptus oil was more toxic to both insect species on filter paper discs and to SZ on maize. The repellent against insects for crude EOs was stronger than that for only main constitutes	Tapondjou et al. (2005)
Hyptis fruti-cosa Salzm., Hyptis pectinata Poit. and Lippia gracilis HBK	Lamiaceae and Verbenaceae	Aedes aegypti	L. gracilis EOs showed strong larvicidal activity due to its carvacrol inside (44.43 %), which caused 100 % larval mortality at 150 ppm	Silva et al. (2008)
Lavandula hybrida, Rosmarinus officinalis and Eucalyptus globulus	Lamiaceae, Myrtaceae	Acanthoscelides obtectus	The larvae showed gradually more tolerant to EOs as they grew older, but they were more sensitive than pupae. Larval mortality increased with the exposure time until 48 h. EOs' vapours were more effective at 10 and 18 °C	Papachristos and Stamopoulos (2002)
Origanum acutidens	Lamiaceae	Sitophilus granarius and Tribolium confusum	EOs respectively showed 68.3 % and 36.7 % mortality against SG and TC adults	Kordali et al. (2008)

(continued)

Table 5.1 (continued)

Source of EOs	Family	Insect (adults, eggs and larvae)	Remarks	References
Pimpinella anisum, Cuminum cyminum, Eucalyptus camaldulensis, Origanum syriacum var. bevanii and Rosmarinus officinalis	Ombelliferae, Myrtaceae and Lamiaceae	Tribolium confusum and Ephestia kuehniella	100 % mortality of the eggs was achieved for anise and cumin EOs, which also showed relatively high LT$_{99}$ value at 98.5 μl/l air. Other EOs performed less active	Tunç et al. (2000)
Carum carvi, Citrus aurantium, Foeniculum vulgare, Thymus vulgaris, Ocimum basilicum, Lavandula angustifolia	Apiaceae, Rutaceae, Lamiaceae	Meligethes aeneus	EOs of C. carvi and T. vulgaris were most efficient considering the mortality (LD$_{50}$ = 197 and 250 μg/cm^2) and repellent effect (Repellent index = 65.6 % and 63.8 %)	Pavela (2011)
Cinnamomum zeylanicum, S. terebinthifolius and E. uvalha	Lauraceae Anacardiaceae Myrtaceae	T. putrescentiaeand and S. pontifica	EOs were more toxic than their major components. All EOs were toxic to the mites, among which a quick lethal response was caused by S. terebinthifolius EOs	De Assis et al. (2011)
Corymbia citriodora, Croton sonderianus, Cymbopogon martini, Lippia alba, Lippia gracilis, Lippia sidoides and Pogostemon cablin	Myrtaceae, Ruphobiaceae, Poaceae, Verbenaceae, Lamiaceae	Nasutitermes corniger	All EOs were toxic to termites. EOs from Pogostemon cablin and Lippia sidoides were the most promising for the control of N. corniger	Lima et al. (2013)

(continued)

Table 5.1 (continued)

Source of EOs	Family	Insect (adults, eggs and larvae)	Remarks	References
Echinophora tenuifolia ssp. *Sibthorpiana, Anethum graveolens* and *Chaerophyllum aromaticum*	*Apiaceae*	*Culex pipiens* L.	EOs of *A. graveolens* displayer the most active with an LC_{50} value of 52.74 mg/L, followed by the EOs of *E. tenuifolia* ssp. *sibthorpiana* and *C. aromaticum* were with an LC_{50} values near 60 mg/L	Evergetis et al. (2013)
Eucalyptus citriodora, Eucalyptus staigeriana, Cymbopogon winterianus and *Foeniculum vulgare*	*Myrtaceae, Poaceae, Apiaceae*	*Callosobruchus maculatus*	The higher the oil concentration, the lower the number of laid eggs and emerged insects. EOs of *EC* and *CW* were repellent to adult and EOs of *FV* were neutral at all concentrations, while EOs of *ES* was neutral (<558 ppm) and repellent at higher concentrations	Gusmão et al. (2013)
Eucalyptus globulus	*Myrtaceae*	*Pediculus humanus capitis*	The major monoterpene 1,8-cineole in EOs revealed toxicity against louse, which the EOs' LT_{50} value was 0.125 mg/cm^2 compared to 0.25 mg/cm^2 of commercial pediculides	Yang et al. (2004)
Melaleuca alternifolia	*Myrtaceae*	*Ceratitis capitata* and *Psyttalia concolor*	EOs showed lower LC_{50} and LD_{50} values for *C. capitata* in all assays (contact: 0.117 µl oil/cm^2 vs. 0.147 µl oil/cm^2; fumigation: 2.239 µl oil/L air vs. 9.348 µl oil/L air; 0.269 % vs.0.638 % of EO, w/w)	Benelli et al. (2013)
Micromeria fruticosa L., *Nepeta racemosa* L. and *Origanum vulgare* L.	*Lamiaceae*	*Tetranychus urticae* Koch and *Bemisia tabaci* Genn	All EOs' vapour induced the highest mortality after 120 h treatment at 2 µl/l air for two pests. *T. urticae* was less sensitive than *B. tabaci* at all doses of EOs for all times	Çalmasur et al. (2006)
Satureja hortensis L., *Ocimum basilicum* L. and *Thymus vulgaris* L.	*Lamiacae*	*Tetranychus urticae* Koch and *Bemisia tabaci* Genn	EOs of *S. hortensis* was found to be the most effective, which caused the highest mortality (>96 %) after 96 h exposure at 3.125 µl/l air	Aslan et al. (2004)

5.4 Conclusions

As summarized previously, a large amount of EOs have insecticidal properties against various species of insects. These properties are related to chemical compositions in complex EOs, the functional groups (alcohols, phenols, etc.) of major compounds and their synergistic effects. It should be noted that these synergies may also involve minor compounds. The EOs represent an attractive potential to substitute synthetic insecticides because of their diversity and efficiency. Besides, it is essential to develop more efficient and green extraction techniques with appreciable yields in order to meet the demand of EOs as natural insecticides one day.

Acknowledgement Authors thank Prof. Chahrazed Boutekedjiret from Ecole Nationale Polytechnique Algeria for her valuable comments and discussion about EO's as insecticides.

References

Aslan I, Özbek H, Çalmasur Ö, Sahín F (2004) Toxicity of essential oil vapours to two greenhouse pests *Tetranychus urticae* Koch and *Bemisia tabaci* Genn. Ind Crop Prod 19:167–173

Baldi I, Lebailly P, Mohammed-Brahim B et al (2003) Neurodegenerative diseases and exposure to pesticides in the elderly. Am J Epidemiol 157:409–414

Benelli G, Canale A, Flamini G et al (2013) Biotoxicity of *Melaleuca alternifolia* (Myrtaceae) essential oil against the Mediterranean fruit fly, *Ceratitis capitata* (Diptera: Tephritidae), and its parasitoid *Psyttalia concolor* (Hymenoptera: Braconidae). Ind Crop Prod 50:596–603

Boutekedjiret C (2009) Essential oils as insecticidal agents. In: Chemat F (ed) Essential oils and aromas: green extraction and applications. Har Krishan Bhalla & Sons, Dehradun, pp 227–247

Çalmasur Ö, Aslan I, Sahin F (2006) Insecticidal and acaricidal effect of three *Lamiaceae* plant essential oils against *Tetranychus urticae* Koch and *Bemisia tabaci* Genn. Ind Crop Prod 23:140–146

Cheng SS, Chua MT, Chang EH et al (2009a) Variations in insecticidal activity and chemical compositions of leaf essential oils from *Cryptomeria japonica* at different ages. Bioresour Technol 100:465–470

Cheng SS, Liu JY, Huang CG et al (2009b) Insecticidal activities of leaf essential oils from *Cinnamomum osmophloeum* against three mosquito species. Bioresour Technol 100:457–464

Daizy RB, Harminder PS, Ravinder KK, Shalinder K (2008) Eucalyptus essential oil as a natural pesticide. Forest Ecol Manag 256:2166–2174

De Assis CPO, Gondim MGC Jr, De Siqueira HAA, Da Câmara CAG (2011) Toxicity of essential oils from plants towards *Tyrophagus putrescentiae* (Schrank) and *Suidasia pontifica* Oudemans (Acari: Astigmata). J Stored Prod Res 47:311–315

Evergetis E, Michaelakis A, Haroutounian SA (2013) Exploitation of Apiaceae family essential oils as potent biopesticides and rich source of phellandrenes. Ind Crop Prod 41:365–370

Fisk AT, Hobson KA, Norstrom RJ (2001) Influence of chemical and biological factors on trophic transfer of persistent organic pollutants in the northwater polynya marine food web. Environ Sci Technol 35:732–738

Garcıa M, Gonzalez-Coloma A, Donadel OJ et al (2007) Insecticidal effects of *Flourensia oolepis* Blake (Asteraceae) essential oil. Biochem Syst and Ecol 35:181–187

Gusmão NMS, de Oliveira JV, Navarro DM do AF et al (2013) Contact and fumigant toxicity and repellency of *Eucalyptus citriodora* Hook., *Eucalyptus staigeriana* F., *Cymbopogon winterianus* Jowitt and *Foeniculum vulgare* Mill. Essential oils in the management of

Callosobruchus maculatus (FABR.) (Coleoptera: Chrysomelidae, Bruchinae). J Stored Prod Res 54: 41–47

Huang Y, Lam SL, Ho SH (2000) Bioactivities of essential oil from *Elletaria cardamomum* (L.) Maton. to *Sitophilus zeamais* Motschulsky and *Tribolium castaneum* (Herbst). J Stored Prod Res 36:107–117

Jemâa JMB, Tersim N, Toudert KT, Khouja ML (2012) Insecticidal activities of essential oils from leaves of *Laurus nobilis* L. from Tunisia, Algeria and Morocco, and comparative chemical composition. J Stored Prod Res 48:97–104

Kéita SM, Vincent C, Schmit JP et al (2001) Efficacy of essential oil of *Ocimum basilicum* L. and *O. gratissimum* L. applied as an insecticidal fumigant and powder to control *Callosobruchus maculatus* (Fab.) [Coleoptera: Bruchidae]. J Stored Prod Res 37:339–349

Ketoh GK, Koumaglo HK, Glitho IA, Huignard J (2006) Comparative effects of *Cymbopogon schoenanthus* essential oil and piperitone on *Callosobruchus maculatus* development. Fitoterapia 77:506–510

Khalfi O, Sahraoui N, Bentahar F, Boutekedjiret C (2008) Chemical composition and insecticidal properties of *Origanum glandulosum* (Desf.) essential oil from Algeria. J Sci Food Agr 88:1562–1566

Kim SI, Park C, Ohh MH et al (2003a) Contact and fumigant activities of aromatic plant extracts and essential oils against *Lasioderma serricorne* (Coleoptera: Anobiidae). J Stored Prod Res 39:11–19

Kim SI, Roh JY, Kim DH et al (2003b) Insecticidal activities of aromatic plant extracts and essential oils against *Sitophilus oryzae* and *Callosobruchus chinensis*. J Stored Prod Res 39:293–303

Kordali S, Cakir A, Ozer H et al (2008) Antifungal, phytotoxic and insecticidal properties of essential oil isolated from Turkish *Origanum acutidens* and its three components, carvacrol, thymol and *p*-cymene. Bioresour Technol 99:8788–8795

Kotan R, Kordali S, Cakir A et al (2008) Antimicrobial and insecticidal activities of essential oil isolated from Turkish *Salvia hydrangea* DC. ex Benth. Biochem Syst Ecol 36:360–368

Lee BH, Annis PC, Tumaalii F, Choi WS (2004) Fumigant toxicity of essential oils from the *Myrtaceae* family and 1,8-cineole against 3 major stored-grain insects. J Stored Prod Res 40:553–564

Lee BH, Choi WS, Lee SE, Park BS (2001) Fumigant toxicity of essential oils and their constituent compounds towards the rice weevil, *Sitophilus oryzae* (L.). Crop Prot 20:317–320

Lemaire G, Terouanne B, Mauvais P et al (2004) Effect of organochlorine pesticides on human androgen receptor activation *in vitro*. Toxicol Appl Pharm 196:235–246

Lima JKA, Albuquerque ELD, Santos ACC et al (2013) Biotoxicity of some plant essential oils against the termite*Nasutitermes corniger* (Isoptera: Termitidae). Ind Crop Prod 47:246–251

Negahbana M, Moharramipour S, Sefidkon F (2007) Fumigant toxicity of essential oil from *Artemisia sieberi* Besser against three stored-product insects. J Stored Prod Res 43:123–128

Ogendo JO, Kostyukovsky M, Ravid U et al (2008) Bioactivity of *Ocimum gratissimum* L. oil and two of its constituents against five insect pests attacking stored food products. J Stored Prod Res 44:328–334

Oliva A, Spira A, Multigner L (2001) Contribution of environmental factors to the risk of male infertility. Hum Reprod 16:1768–1776

Papachristos DP, Stamopoulos DC (2002) Toxicity of vapours of three essential oils to the immature stages of *Acanthoscelides obtectus* (Say) (Coleoptera: Bruchidae). J Stored Prod Res 38:365–373

Park IK, Lee SG, Choi DH et al (2003) Insecticidal activities of constituents identified in the essential oil from leaves of *Chamaecyparis obtusa* against *Callosobruchus chinensis* (L.) and *Sitophilus oryzae* (L.). J Stored Prod Res 39:375–384

Pavela R (2011) Insecticidal and repellent activity of selected essential oils against of the pollen beetle, *Meligethes aeneus* (Fabricius) adults. Ind Crop Prod 34:888–892

Prajapati V, Tripathi AK, Aggarwal KK, Khanuja SPS (2005) Insecticidal, repellent and oviposition-deterrent activity of selected essential oils against *Anopheles stephensi, Aedes aegypti* and *Culex quinquefasciatus*. Bioresour Technol 96:1749–1757

Rozman V, Kalinovic I, Korunic Z (2007) Toxicity of naturally occurring compounds of *Lamiaceae* and *Lauraceae* to three stored-product insects. J Stored Prod Res 43:349–355

Saidj F, Rezzoug SA, Bentahar F, Boutekedjiret C (2008) Chemical composition and insecticidal properties of *Thymus numidicus* (Poiret) essential oil from Algeria. J Essent Oil Bear Pl 11:397–405

Saiyed H, Dewan A, Bhatnagar V et al (2003) Effect of endosulfan on male reproductive development. Environ Health Perspect 11:1958–1962

Seri-Kouassi BP, Kanko C, Aboua LRN et al (2004) Action des huiles essentielles de deux plantes aromatiques de Côte-d'Ivoire sur *Callosobruchus maculatus* F. du niébé. C R Chim 7:1043–1046

Silva WJ, Doria GAA, Maia RT et al (2008) Effects of essential oils on *Aedes aegypti* larvae: Alternatives to environmentally safe insecticides. Bioresour Technol 99:3251–3255

Singh HP, Batish DR, Setia N, Kohli RK (2005) Herbicidal activity of volatile oils from *Eucalyptus citriodora* against *Parthenium hysterophorus*. Ann Appl Biol 146:89–94

Tapondjou AL, Adler C, Fontem DA et al (2005) Bioactivities of cymol and essential oils of *Cupressus sempervirens* and *Eucalyptus saligna* against *Sitophilus zeamais* Motschulsky and *Tribolium confusum* du Val. J Stored Prod Res 41:91–102

Tapondjou LA, Adler C, Bouda H, Fontem DA (2002) Efficacy of powder and essential oil from *Chenopodium ambrosioides* leaves as post-harvest grain protectants against six-stored product beetles. J Stored Prod Res 38:395–402

Tunç I, Berger BM, Erler F, Dagli F (2000) Ovicidal activity of essential oils from five plants against two stored-product insects. J Stored Prod Res 36:161–168

Yang YC, Choi HY, Choi WS et al (2004) Ovicidal and adulticidal activity of Eucalyptus globulus leaf oil terpenoids againstPediculus humanus capitis (Anoplura: Pediculidae). J Agric Food Chem 52:2507–2511

Chapter 6
Essential Oils as Green Solvents

Abstract This chapter introduces the solvency of essential oils (EOs) as green solvents, in particular, application of terpenes as solvents in extraction. The physiochemical properties of such solvents are predicted in comparison to the conventional petroleum-based solvents. Several case studies for different purposes are provided for future consideration.

Keywords Essential oils · Green solvents · Physiochemical properties · Application

As described earlier in the first chapter, essential oils (EOs) are generally odorous mixtures of many terpene hydrocarbons formed by plant metabolism. The common EOs produced by conventional methods have been on the rise in quantity over the years, in which monoterpene hydrocarbons (e.g. d-Limonene, α-pinene, *p*-cymene) accounted for a large proportion. These EOs can be considered a natural resource of terpenes which may appeal to many sectors including agri-food, cosmetic, pharmaceutical, perfumery and fragrance industry. For instance, limonene as the major terpenes (91–97 %) in citrus EOs, can be obtained through distillation of by-products in the production of citrus juices. Food industry would like to valorise such natural co-products which can be considered for other purposes such as industrial solvent to obtain food grade ingredient such as colours, antioxidants, etc. The conventional solvents such as n-hexane generally used in industry, are derived from petroleum or halogenated solvents with a no doubt negative effect on both human health and environment. In addition, the industries are encouraged to use alternative and eco-friendly solvents by governments and strict legislations or standards with the consideration of both technical feasible and economically viable aspects.

Terpenes are natural solvents derived from EOs, which have represented their potential to be alternatives to petroleum-based solvents in various industrial applications. They differ in physical properties due to acyclic, bicyclic or monocyclic structures. d-Limonene is a biodegradable, low-toxic and cost-effective terpene extracted from citrus peels as agricultural waste. A growing interest in various application of

Y. Li et al., *Essential Oils as Reagents in Green Chemistry*, SpringerBriefs in Green Chemistry for Sustainability, DOI 10.1007/978-3-319-08449-7_6

this renewable reagent has been taken into account as its cleaning and degreasing properties were recognized with a considerable performance (Toplisek and Gustafson 1995; Chemat et al. 2012). Recently, the extraction of oil from oleiferous materials using d-Limonene as an alternative to organic solvents has been investigated (Mamidipally and Liu 2004; Virot et al. 2008a). The quantitative and qualitative analysis of terpene extracts presented similar results to those obtained using n-hexane. Alpha-pinene is another interesting potential solvent, which are generally obtained by distillation of pine oleoresins or fractionation of steam-distilled wood turpentine. It represents as the major constituent in turpentine oils from most conifers, or as one of components in wood, bark or leaf oils from a wide variety of other plants such as rosemary, basil, rose, etc. p-Cymene present widely in tree leaf oils has been used as either a solvent for dyes and varnishes, or an additive in fragrances and musk perfumes, or an odour masking agent for industrial products.

6.1 The Evaluation of Physiochemical Properties

The solvent power of these compounds can be evaluated by prediction of their physical and thermodynamic properties using either conventional or modern methods. The Hansen solubility parameters (HSP) theory has partitioned the fundamental Hildebrand's energy into three distinctive forms of energy in terms of dispersion, polar and hydrogen bond forces (Abbott et al. 2013). In general, the HSP follow the classical "like dissolve like" rule that the smaller the dissimilarity of the HSP distance between molecules, the greater the affinity between them. There are six HSP algorithms in HSPiP software, among which methods of Yamamoto and Stefanis-Panayiotou (S-P) are highly recommended. The S-P method is recognized as the most extensive and accurate group-contribution method to determine the four HSP values (δ_{tot}, δ_d, δ_p and δ_h) of molecules. It breaks down the molecular structure into first- and second-order functional groups, which respectively describe the overall structure of the molecule corresponding to UNIFAC groups and improve the description of the molecular structure for higher prediction accuracy (Stefanis and Panayiotou 2008). The Yamamoto-Molecular Break (Y-MB) method provides an automatic way of creating only first-order groups through simplified molecular input line entry syntax (SMILES) or 3D molecule input. The combination of an adaptive neural network methodology gives the best predictive power for HSP, which inter-group interactions automatically get fitted by the relative strengths of the neural interconnections. Moreover, it allows the evaluation of other parameters (e.g. boiling point, density, molecular volume, etc.) which other methods cannot achieve. Therefore, the theoretical solubility of main components in EOs, relevant common solvents and interesting solutes has been calculated by Y-MB method through their chemical structures (Fig. 6.1).

Instead of 3D sphere of Hansen solubility parameters, this 2D solubility diagram (δ_h vs. δ_p) excluding the insignificantly similar δ_d, can help to better visualize the polarity of components so as to further interpret the dissolving mechanism. For

Fig. 6.1 The predicted
Hansen solubility parameters
of relevant solutes and
solvents

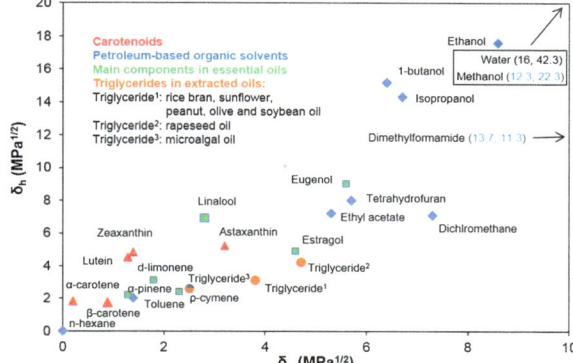

instance, terpenes used in previous experimental studies are closer to carotenoids and oil triglycerides than other organic solvents, which prove their potency of being alternatives to other common solvents. From a physiochemical point of view, terpenes are lipophilic solvents that can dissolve in organic solvents but are practically insoluble in water. In addition to the similar HSP values to n-hexane, terpenes with higher dielectric constant have higher polarity and dissociating power than n-hexane. Besides, their higher flash points signify less flammable and less hazardous. It is worth to point out that this method is still under development which has shown its superiority with more convincing results. In order to get rational and accurate HSP values at present, it is also insightful to compare the HSP predictions of these two above-mentioned methods because they both have their strengths and limitations.

The COnductor-like Screening MOdel for Real Solvents (COSMO-RS) is one of the most popular continuum solvation models based on mono-molecular quantum chemical computations that generate a priori descriptors deriving from molecular structure information (Klamt et al. 2010). The σ-profile and σ-potential in COSMO-RS theories, as well as other thermodynamic parameters, can help to pre-screening and to classify various solvents with the help of multivariate statistical analyses so as to select appropriate solvents for different applications (Moity et al. 2012). Compared to conventional group contribution methods, this modern approach can well distinguish between stereoisomers though their real molecular structures. Moreover, it can well interpret interactions of complex fluid systems at a desired temperature or pressure, including multiphase mixture and macromolecules.

6.2 Main Purposeful Applications

Oils and fats from animal or vegetable origin are extracted with nonpolar solvents such as n-hexane, which is a commonly-used, petroleum-derived organic solvent with a relatively low boiling point (69 °C). It has been selected as an excellent

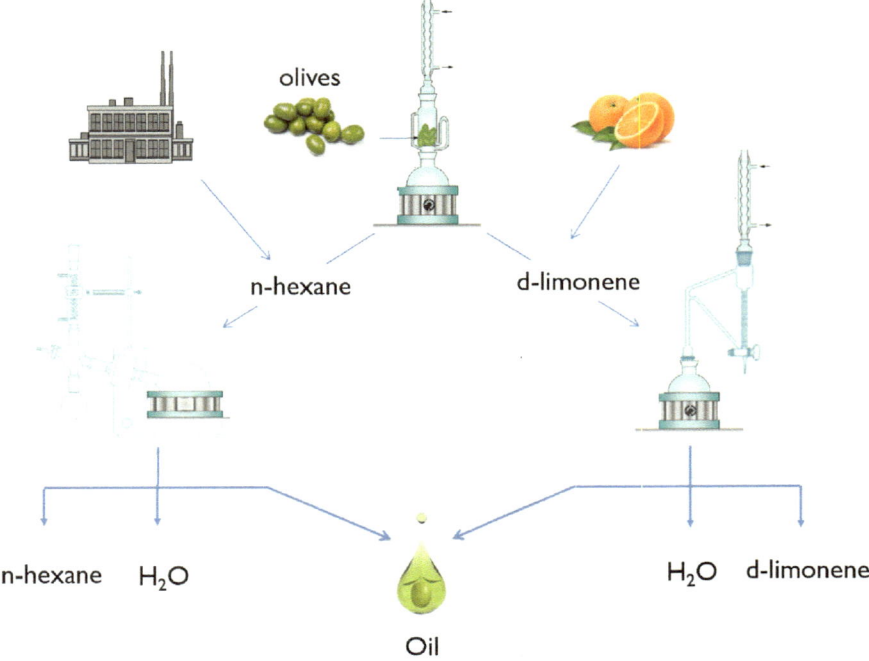

Fig. 6.2 The concept of oil extraction using n-hexane and d-limonene as the solvent

solvent in classic Soxhlet extraction for many years because of its main advantages of high efficiency and simple recovery. However, since its negative impacts on human health and environment have been identified, questions of using n-hexane arise due to its leakage during extraction and recovery. Thus, scientists and industries are required to explore alternative solvents to n-hexane in order to reduce its emissions and adverse impacts. Early works on extraction of vegetable oils using d-Limonene as the solvent have been reported with favourable comments on yield, quality and scale-up (Liu and Mamidipally 2005). Due to the high boiling point of terpenes, the Clevenger apparatus coupling to a cohobation system allows a continuous azeotropic distillation for the removal of terpenes from a mixture of oils and water (Fig. 6.2). The higher energy consumption in terpene recovery during extraction is also the main drawback of using such solvents, as well as their higher viscosity and density. Furthermore, other terpenes like α-pinene and p-cymene have recently attempted to achieve an efficient oil extraction. Bertouche et al. (2013) proved the potential of using α-pinene as a good alternative to n-hexane with comparable results. However, p-cymene among terpene solvents showed the most promising performance for substitution of n-hexane and alcoholic solvents (ethanol, butanol and isopropanol) respectively in terms of extraction yield and selectivity (Dejoye-Tanzi et al. 2012; Li et al. 2014). Moreover, the high temperature for boiling terpene can decrease the viscosity and thus facilitate a better diffusion in order to get a better lipid yield. The green extraction process has also been developed with desired results through combination of green extraction techniques and proved terpene solvents, which can significantly reduce treatment time and energy (Virot et al. 2008b; Dejoye-Tanzi et al. 2013).

The food moisture is often determined by oven drying methods, which are limited for matrices containing a high level of volatile compounds due to the overestimation of actual water contents in food samples. Hence, the distillation methods have been recognized as the most suitable methods, among which the continuous and refluxing Dean-Stark distillation became the reference method for water determination in food products and subsequently in herbs and spices (Balladin and Headley 1999; Brunnemann et al. 2002; Fleury et al. 2006). The recommended solvent for Dean-Stark distillation is toluene, which can be obtained from the petroleum industry and is commonly used in chemical synthesis or in cosmetic and pharmaceutical industries as an extraction solvent. However, like in the case of n-hexane, such solvent has to be gradually eliminated due to increasing environmental and health concerns. Veillet et al. (2010) has replaced toluene by d-Limonene in the Dean-Stark procedure for moisture determination in food products. Although the boiling point of d-Limonene (176 °C) is higher than that of toluene (111 °C), the azeotropic distillation depending on the ability of solvent (d-limonene) can form an azeotropic mixture at 97.4 °C with in situ water contained in the food matrix, which has proved potential application of limonene as an alternative to toluene. The azeotropic distillation kinetics appeared similar with minor variations. Although the water recovery in the beginning was laggard when d-Limonene was used, the total time required to achieve complete water recovery was shorter than that of toluene. Therefore, the excess energy consumption can be compensated by the shorter processing time. Bertouche et al. (2012) obtained similar kinetics with α-pinene as an alternative solvent in moisture determination of food products. The new Deak-Stark procedure using terpenes as solvents has proved to be available for a wide range of food products, including vegetables, herbs, spices and meat. It has been found that d-Limonene showed a better performance than that of α-pinene, especially for moisture in meat.

The traditional extraction of food colors such as carotenoids is usually carried out using organic solvents such as dichloromethane, acetone, chloroform, etc. The volatility and the dissolving power of these petroleum-origin solvents make themselves to be preferential for an efficient application. However, these solvents are harmful to human and environment. Moreover, the solvent residues may contaminate the end-products. As a consequence, a greener extraction of lycopene from fresh tomatoes using d-Limonene instead of dichloromethane has been proposed (Chemat-Djenni et al. 2010) (Fig. 6.3). The d-Limonene showed its potential value of protecting environment through reduction of volatile organic compounds even though dichloromethane showed a higher extraction yield. This proposed approach explores a new solution for the post-processing of by-products in industry.

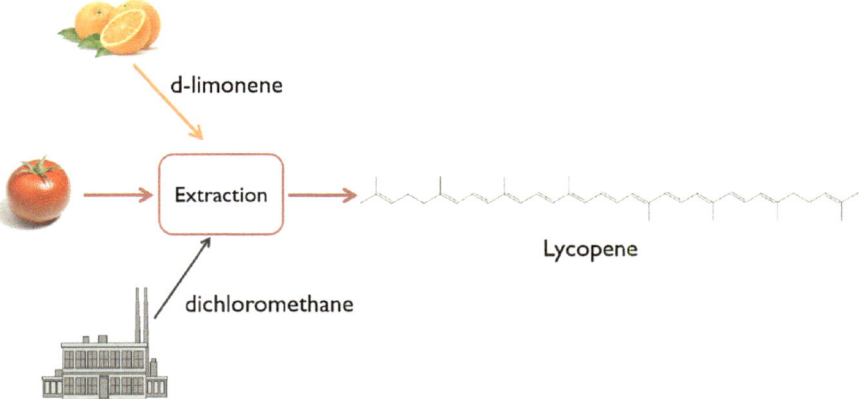

Fig. 6.3 The lycopene extraction using dichloromethane and d-limonene as the solvent

6.3 Conclusions

The major monoterpenes in EOs have proved their potency of becoming alternatives to petroleum-based solvents in various applications, which are mostly in line with the 5th principle of green chemistry and green extraction concept. The procedure using such novel solvents typically involves less energy, less hazardous substances and renewable solvents, which lead to a real sustainable process with integration of authorized green techniques. Using terpenes as alternative solvents is of great interest for future extraction of natural bioactive compounds in food, cosmetic and medical field.

References

Abbott S, Hansen C, Yamamoto H (2013) The minimum possible theory. In: Hansen solubility parameters in practice, Hansen-Solubility.com, pp 9–13
Balladin DA, Headley O (1999) Evaluation of solar dried thyme (Thymus vulgaris Linne) herbs. Renew Energ 17:523–531
Bertouche S, Tomao V, Hellal A, Boutekedjiret C, Chemat F (2013) First approach on edible oil determination in oilseeds products using alpha-pinene. J Essent Oil Res 25:439–443
Bertouche S, Tomao V, Hellal A, Ruiz K, Hellal A, Boutekedjiret C, Chemat F (2012) First approach on moisture determination using alpha-pinene. Food Chem 134:602–605
Brunnemann KD, Qi J, Hoffmann D (2002) Chemical profile of two types of oral snuff tobacco. Food Chem Toxicol 11:1699–1703
Chemat S, Tomao V, Chemat F (2012) Limonene as green solvent for extraction of natural products. In: Mohammad A (ed) Green solvents Properties and applications in chemistry. Springer, Berlin, p 177
Chemat-Djenni Z, Ferhat MA, Tomao V, Chemat F (2010) Carotenoid extraction from tomato using a green solvent resulting from orange processing waste. J Essent Oil Bearing Plant 13:139–147

Dejoye-Tanzi C, Abert-Vian M, Chemat F (2013) New procedure for extraction of algal lipids from wet biomass: a green clean and scalable process. Bioresour Technol 134:271–275

Dejoye-Tanzi C, Abert-Vian M, Ginies C, Elmaataoui M, Chemat F (2012) Terpenes as green solvents for extraction of oil from microalgae. Molecules 17:8196–8205

Fleury M, Boyd D, Al-Nayadi K (2006) Water saturation from NMR, resistivity and oil-base core in a heterogeneous Middle-East carbonate reservoir. Petrophysics 47:60–73

Klamt A, Eckert F, Arlt W (2010) COSMO-RS: an alternative to simulation for calculating thermodynamic properties of liquid mixtures. Annu Rev Chem Biomol Eng 1:101–122

Li Y, Fine F, Fabiano-Tixier AS, Abert-Vian M, Carre P, Pages X, Chemat F (2014) Evaluation of alternative solvents for improvement of oil extraction from rapeseeds. C R Chim 17:242–251

Liu SX, Mamidipally PK (2005) Quality comparison of rice bran oil extracted with d-limonene and hexane. Cereal Chem 82:209–215

Mamidipally PK, Liu SX (2004) First approach on rice bran oil extraction using limonene. Eur J Lipid Sci Tech 106:122–125

Moity L, Duran M, Benazzouz A (2012) Panorama of sustainable solvents using the OCSMO-RS approach. Green Chem 14:1132–1145

Stefanis E, Panayiotou C (2008) Prediction of Hansen solubility parameters with a new group-contribution method. Int J Thermophys 29:568–585

Toplisek T, Gustafson R (1995) Cleaning with d-limonenes: a substitute for chlorinated solvents. Precis Clean 3:17–22

Veillet S, Tomao V, Ruiz K, Chemat F (2010) Green procedure using limonene in the Dean-Stark apparatus for moisture determination in food products. Anal Chim Acta 674:49–52

Virot M, Tomao V, Chinies C, Chemat F (2008a) Total lipid extraction of food using d-limonene as an alternative to n-hexane. Chromatographia 68:311–313

Virot M, Tomao V, Chinies C, Visinoni F, Chemat F (2008b) Green procedure with a green solvent for fats and oils' determination. J Chromatogr A 1196–1197:147–152

Chapter 7
Essential Oils as Synthons for Green Chemistry

Abstract This chapter briefly presents a historical development of organic synthesis first and the application of main ingredients in essential oils (EOs) as synthons in various purposeful syntheses has been provided afterwards for a sustainable future, which is based on the principle of green chemistry.

Keywords Essential oils · Ingredients · Synthesis · Green chemistry

Due to the fact that the starting materials used for current organic synthesis are mainly petroleum-derived substances, natural components of plant origin have attracted more attention in greener syntheses for a sustainable chemistry. Essential oils (EOs) consist of diverse compounds, which can be isolated by distillation for potential use as renewable building blocks in the synthesis of basic or fine chemicals. As regards the complex composition of EOs, the volatile fraction (e.g. linalool, eugenol, limonene, etc.) of these secondary metabolites is essentially terpenoid- or phenolic-origin, which mostly has a smaller molecular weight (<350 g/mol). These compounds with different unsaturated levels contain functional or asymmetric carbon groups, which can be used as starting materials for functional modifications in purposeful syntheses.

7.1 Development of the Chemistry of Terpene

Numerous organic syntheses have been carried out since the first urea synthesis happened in the early of 19th century. This significant progress can help not only to develop the synthetic methodology and structural elucidation, but also to understand the reaction mechanisms and enantioselectivity so as to make use of them better. Nevertheless, the majority of starting materials or carbon substrates, which have been used in organic synthesis nowadays, is petrochemicals. Subsequently, chemists have attempted to use biomass as renewable source of starting materials

© The Author(s) 2014   63
Y. Li et al., *Essential Oils as Reagents in Green Chemistry*, SpringerBriefs in Green Chemistry for Sustainability, DOI 10.1007/978-3-319-08449-7_7

in the face of gradually depleted fossil resource. The improvement of syntheses which mainly focus on the yield of desired products has been lasted until the famous twelve principles of green chemistry have been enacted in the late 20th century (Anastas and Warner 1998). The natural substances such as those found in EOs should be considered in syntheses to ease the energy dilemma, which is aligned with the 7th principle of green chemistry. For instance, two chiral molecules (shikimic and quinic acid) from biomass have been proved to facilitate the hemisynthesis of an achiral molecule (tamiflu). In the meanwhile, new syntheses such as enzymatic modification of natural origin molecules have been recently developed (Groussin and Antoniotti 2012).

7.2 Essential Oils as Ingredients in Synthesis

In general, products in industrial chemistry are distinguished with two categories. The category of commodity includes low value products with simple structures that usually need for mass production according to the automated and optimized continuous process, while the category of fine chemicals involves high value-added products with sophisticated performances which are typically produced in smaller quantities. The renewable main ingredients in EOs can not only be reserved for the products in the first category but they can also be synthons for syntheses of products in the second category. The interest of using them as building blocks for industrial chemical products has been put into practice in various applications, some of which have been summarized in Table 7.1. It is worth noting that some of such synthons can also transform in between reciprocally. For instance, δ-carene and α-pinene can be transformed into dipentene (limonene), which the latter can be transformed into cymene using various catalytic systems.

Table 7.1 Main interesting syntheses using components from essential oils as synthons

Main synthons	Essential oils	Main target products	References
α-Pinene β-Pinene 3-carene	Turpentine (*Pinus sp.*)	 Myrcene, terpenyl ethers, *cis*-pinane, camphor, dipentene, myrac aldehyde, campholenic aldehyde	Saudan (2007); Lemchko et al. (2007); Suh et al. (2003); Swift (2004)

(continued)

Table 7.1 (continued)

Main synthons	Essential oils	Main target products	References
Limonene	Citrus (*Citrus sp.*)	8-thio-*p*-menthene, carvol, cymene, terpenyl ester	Tani et al. (1984); Naigre et al. (1994); Weyrich and Hoelderich (1997)
Citral	Lemongrass (*Cymbopogon citratus*)	Retinol (vitamin A), α-ionones	Andriamialisoa et al. (1993); Moronkola et al. (2005)
Valencene	Valencia oranges (*Citrus sinensis*)	Nootkatone	Furusawa et al. (2005); Dubal et al. (2008)

(continued)

Table 7.1 (continued)

Main synthons	Essential oils	Main target products	References
Pulegone (Figure 1)	Pennyroyal (*Mentha piperita*)	8-thio-*p*-menthan-3-one, valerenic acid, (+)-menthofuran, methanol Nepetalactone	Dia et al. (2010); Kopp et al. (2009); Chakraborty and Chattopadhyay (2008)
Nepetalactone	Catnip (*Nepeta cataria*)	Intermediate to (+)-englerin A, iridodial	Chauhan et al. (2004); Willot et al. (2009)
Eugenol	Clove (*Syzygium aromaticum*)	Intermediate to (-)-platensimycin, vanillin	Eey and Lear (2010); Luu et al. (2009)

(continued)

Table 7.1 (continued)

Main synthons	Essential oils	Main target products	References
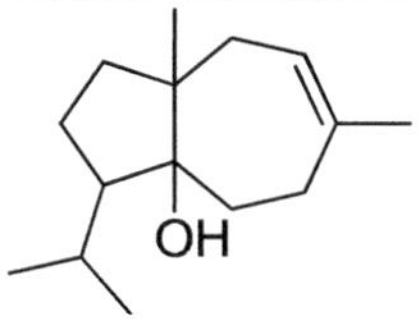 Carotol	Carrot seeds (*Daucus carota*)	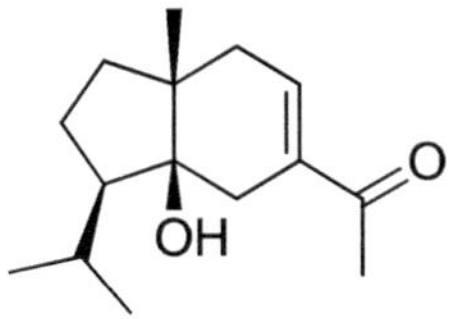 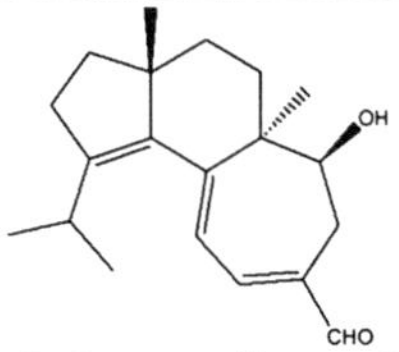 Intermediate to cyathane skeleton, allocyathine B2	Kula et al. (2002); Bonikowski et al. (2009)
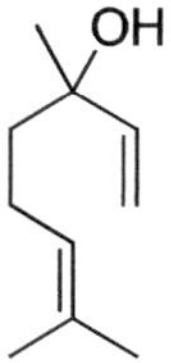 Linalool (Figure 2)	Coriander (*Coriandrum sativum*)	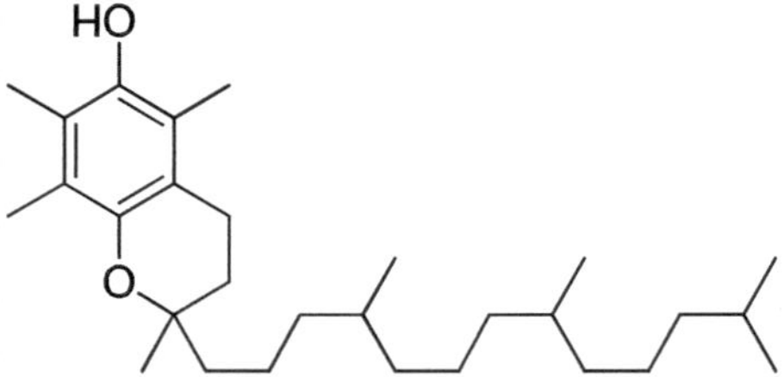 α-tocopherol (vitaminE)	Gembus et al. (2009)

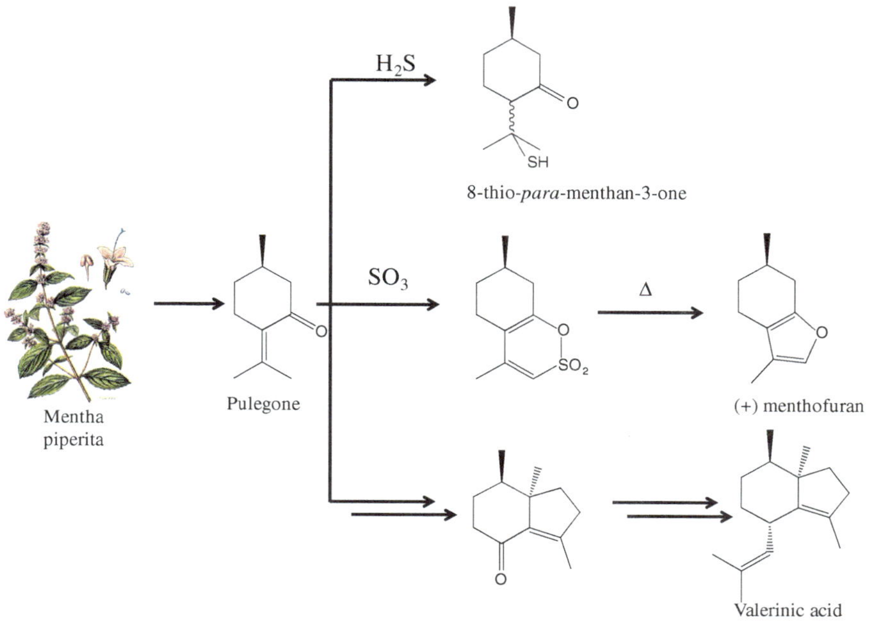

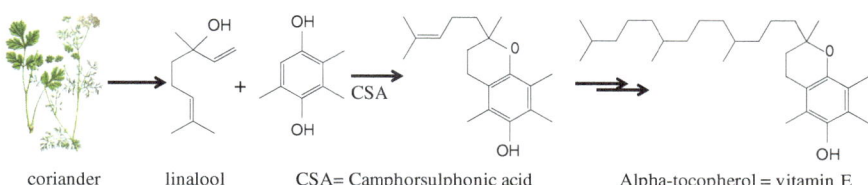

Fig. 7.1   Pulegone from *Mentha piperita* as a synthon for different syntheses

Fig. 7.2   The conversion of linalool from coriander to α-tocopherol

7.3   Conclusions

As can be learned through various applications above, the interest of using major compounds in EOs as synthons in organic synthesis has represented a signifi-cant possibility and success. These natural synthons from renewable biomass are expected to play an important role in the future syntheses of industrial chemicals. Meanwhile, other less volatile or less refined compounds in EOs which are read-ily available should be emphasized to study their potential in the total synthesis as starting materials. Besides, with the aim of maximizing production of interesting metabolites for purposeful syntheses, the relevant technology of gene combination for plant metabolisms should also be considered.

A scanning electron micrograph of untreated caraway seed

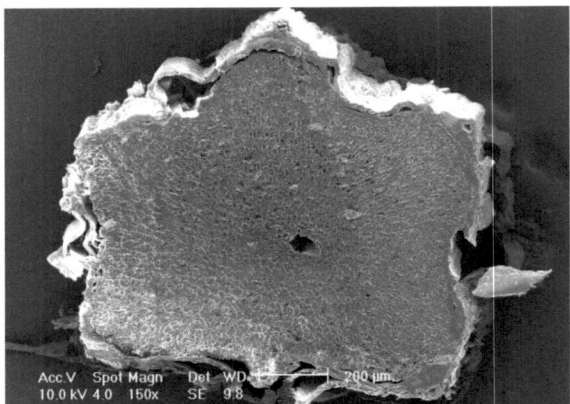

References

Anastas PT, Warner JC (1998) Green chemistry: theory and practice. Oxford University Press, New York, p 30

Andriamialisoa Z, Valla A, Zennache S, Giraud M, Potier P (1993) New preparation of an important synthon for vitamin A synthesis. Tetrahedron Lett 34:8091–8092

Bonikowski R, Kula J, Bujacz A, Wajs A, Majzner W (2009) Natural (+)-carotol as a donor of chirality in the synthesis of enantiomerically pure hydroindene building blocks. Tetrahedron: Asymmetr 20:2583–2588

Chakraborty A, Chattopadhyay S (2008) Stimulation of menthol production in *Mentha piperita* cell culture. In Vitro Cell Dev Biol Plant 44:518–524

Chauhan RK, Zhang QH, Aldrich RJ (2004) Iridodials: enantiospecific synthesis and stereochemical assignment of the pheromone for the golden-eyed lacewing, *Chrysopa oculata*. Tetrahedron Lett 45:3339–3340

Dia RM, Belaqziz R, Romane A, Antoniotti S, Duñach E (2010) Flavouring and odorant thiols from renewable resources by InIII-catalysed hydrothioacetylation and lipase-catalysed solvolysis. Tetrahedron Lett 51:2164–2167

Dubal SA, Yogesh TP, Momin SA (2008) Biotechnological routes in flavour industries. Adv Biotech 6:20–31

Eey STC, Lear MJ (2010) A bismuth(III)-catalyzed Friedel-Crafts cyclization and stereocontrolled organocatalytic approach to (−)-platensimycin. Org Lett 12:5510–5513

Furusawa M, Hashimoto T, Noma Y, Asakawa Y (2005) Highly efficient production of nootkatone, the grapefruit aroma from valencene, by biotransformation. Chem Pharm Bull 53:1513–1514

Gembus V, Sala-Jung N, Uguen D (2009) A convenient access to (all-rac)-α-tocopherol acetate from linalool and dihydromyrcene. Bull Chem Soc Jpn 82:829–842

Groussin AL, Antoniotti S (2012) Valuable chemicals by the enzymatic modification of molecules of natural origin: terpenoids, steroids, phenolics and related compounds. Bioresour Technol 115:237–243

Kopp S, Schweizer WB, Altmann KH (2009) Total synthesis of valerenic acid. Synlett 2009:1769–1772

Kula J, Bonikowski R, Staniszewska M et al (2002) Transformation of carotol into the hydroindane-derived musk odorant. Eur J Org Chem 11:1826–1829

Lemechko P, Grau F, Antoniotti S, Duñach E (2007) Hydroalkoxylation of non-activated olefins catalysed by Lewis superacids in alcoholic solvents: an eco-friendly reaction. Tetrahedron Lett 48:5731–5734

Luu TXT, Lam TT, Le TN, Duus F (2009) Fast and green microwave-assisted conversion of essential oil allylbenzenes into the corresponding aldehydes via alkene isomerization and subsequent potassium permanganate promoted oxidative alkene group cleavage. Molecules 14:3411–3424

Moronkola DO, Aiyelaagbe OO, Ekundayo O (2005) Syntheses of eight fragrant terpenoids [ionone derivatives] via the aldol condensation of citral and eight ketones. J Essent Oil Bear Pl 8:87–98

Naigre R, Chenal T, Cipres I, Kalck P (1994) Carbon monoxide as a building block in organic synthesis. Part V. Involvement of palladium-hydride species in carbonylation reactions of monoterpenes. X-ray crystal structure of $[Ph_3PCH_2CH\ CHPh]_4[PdCl_6][SnCl_6]$. J Organomet Chem 480:91–102

Saudan LA (2007) Hydrogenation processes in the synthesis of perfumery ingredients. Acc Chem Res 40:1309–1319

Suh YW, Kim NK, Ahn WS, Rhee HK (2003) One-pot synthesis of campholenic aldehyde from α-pinene over Ti-HMS catalyst II: effects of reaction conditions. Ind Crop Prod 50:596–603

Swift KAD (2004) Catalytic transformation of the major terpene feedstocks. Top Catal 27:143–155

Tani KY, Akutagawa T, Kumobayashi S et al (1984) Highly enantioselective isomerization of prochiral allylamines catalyzed by chiral diphosphine rhodium (I) complexes. Preparation of optically active enamines. J Am Chem Soc 106:5208–5217

Weyrich PA, Hoelderich WF (1997) Dehydrogenation of α-limonene over Ce promoted, zeolite supported Pd catalysts. Appl Catal A: Gen 158:145–162

Willot M, Radtke L, Könning D et al (2009) Total synthesis and absolute configuration of the guaiane sesquiterpene englerin A. Angew Chem Int Ed 48:9105–9108